HERMETICA TRIPTYCHA VOLUME I

SCRIBE SANGUINE QUIA SANGUIS SPIRITUS

Hermetica Triptycha

VOLUME I
THE MERCURY ELEMENTAL YEAR
WITH EPHEMERIDES, 1925–2050

by
Gary P. Caton

The first-ever comprehensive and integral treatment of Mercury's retrogrades, with a one hundred and twenty-five year ephemeris to track the sequential pattern of the planet Mercury's backward trickster medicine dance as it cycles through: three repeated degrees over forty days, three signs of one elemental triplicity each year, the four elements every six to seven years, twelve elemental returns over seventy nine years.

With additional analysis, lessons, and exercises including: the magical power of the image derived from Mercury's cycle as a visible "star," the philosophical, psycho-spiritual and practical ramifications of our views and attitudes toward Mercury retrograde, and individual chapters for using each of the four cycles of Mercury's retrogrades to facilitate personal growth, psycho-spiritual transformation, and real-world magic.

RUBEDO PRESS
AUCKLAND
2017

© Gary P. Caton 2017.
All rights reserved. No part of this work may be reproduced without express permission from the publisher. Brief passages may be cited by way of criticism, scholarship, or review, as long as full acknowledgement is given.

Although every precaution has been taken to verify the accuracy of the information contained herein, the authors and publisher assume no responsibility for any errors or omissions. This book is not intended as a substitute for medical advice. No liability is assumed for damages that may result from the use of information contained within.

Hermetica Triptycha Volume I
The Mercury Elemental Year
with Ephemerides, 1925–2050
by Gary P. Caton

Published by Rubedo Press
Auckland

Rubedo Press
132 Lone Kauri Road
Auckland 0772
NEW ZEALAND

www.rubedo.press

Book design and typesetting by Joseph Uccello.

Interior illustrations to Section I by Aaron Cheak; illustrations to Sections II–III by Joseph Uccello, unless otherwise noted.

Cover design by Aaron Cheak, featuring the "Second Key" of Basil Valentine's *Zwölf Schlussel* (reproduced after the edition of Eugène Canseliet).

ISBN 978-0-473-41688-1

First published by Rubedo Press in 2017.

To the eternal spirit of Hermes,
in many Divine forms
as friend to humanity.

especially as *Hegetor Oneiron*,
Bringer of Dreams, through the gate of polished horn,
to be accomplished by those who see them.

And to all those who have served well
as *tetelesmenoi Hermei*,
initiates into the Mysteries of Hermes.

TABLE OF CONTENTS

Preface 9
Introduction: A User's Guide 13

SECTION I
The Mercury Elemental Year in the Birth Chart

Chapter I: The Astronomy of the Mercury Elemental Year 21
Chapter II: Delineating the Mercury Elemental Year 31

SECTION II
The Mercury Elemental Year as a Transformative Process

Chapter III: Animism, Magic, and Myth 93
Chapter IV: The Alchemy of Transformation 105
Chapter V: Elemental Magic 135
Chapter VI: The Twelve Elemental Returns 151

SECTION III
Appendix: Mercury Elemental Year Ephemeris 183

Notes 232
Bibliography 238

Preface

The minstrel boy will understand,
He holds a promise in his hand.
He talks of better days ahead,
And by his words your fortune's read.

—GORDON LIGHTFOOT,
"MINSTREL OF THE DAWN"

IT'S 3AM, AND suddenly I'm awakened, not by an alarm or noise or even a dream, but by a deeper, primal urge—to bear witness as the world begins itself anew. I rise quickly, excited with the anticipation and curiosity of a child on Christmas morning. Making my way out to the beach, I look up to see the great Milky Way stretched out across the sky, like a grand flowing garland of tiny star flowers.

For the ancient Egyptians the Milky Way was Nut, great goddess of the night sky. I beckon my trained eyes to seek out the shape of her arched body in its primordial yoga pose, and soon I can see her smiling down upon me. I say a prayer that I may be worthy of the magic that I am about to attempt and ask that she grant me audience with the great Thoth,—he who once eased Nut's primordial birth pangs by outwitting her controlling son Ra, the sun god.

You see, when Ra discovered that his mother was about to give birth again, like a jealous child, he forbade it, and decreed she could not give birth any of the 360 days of the Egyptian year. Eventually Ra's plan was foiled by Thoth, who created five extra days by besting the moon god Khonsu in a game of dice. On these five days Nut gave birth to Horus, Oriris, Set, Isis, and Nepthys.

Thoth not only helped to midwife the Egyptian pantheon, he was also known as the inventor of writing, magic, and astrology. Later, the Greeks equated Thoth with their Hermes, who in turn became Mercury for the Romans. So it is on this morning of February 24, 2015 that I seek the blessing of the wandering star known as

Mercury, for my own new creations. Accordingly, I have arranged for the inaugural Sky Astrology Conference to culminate at a time when Mercury is most visible.

Standing on a beach in the Bahamas during the darkest hour before dawn, I hope to capture a photograph of the most elusive of the visible planets, that it may serve as a talisman to carry forward the magic of this new event. As the first glow arrives on the eastern horizon, I am ready. Soon I see him, and as Mercury rises and brightens he casts a long glimmering reflection across the orange and purple waters, streaking toward the herma—the mound of stones that marks the crossroads between our classroom and the beach.

These sacred mounds have been erected for millennia by travelers in honor of the ancient god of highways, and I have prayed by them here for days, asking for clear weather and for Mercury to show himself to us. Now, my prayers have been answered. It has been the most divinely inspired event of my career thus far, and I feel the presence of thousands of years of reverence coursing through me as the day breaks—dividing the night from day and beginning the world again. To confirm that I am indeed carrying the magic of Hermes with me, I am treated to a final enchanted encounter on the road to the airport, as Thoth arrives to bid adieu, in the form of his sacred bird, the ibis.

Since that time, both the vision and substance of this work have been greatly broadened and deepened into this final form. I cannot help but believe this is because the spirit of Hermes flows through it. It is my hope that this volume may serve to invoke and activate the same magical capacities for creating meaningful moments and works in the reader. By attuning ourselves to the cosmic rhythms of the spirit of Hermes and ritually arranging the elements of our lives in concert, I believe we can conjure the deepest magic of co-creation. As the Gertrude Stein character in Woody Allen's movie *Midnight in Paris* advises: "We all fear death and question our place in the universe. The artist's job is not to succumb to despair, but to find an antidote for the emptiness of existence."[1] I beckon the reader to imagine this volume as a guide for the preparation of such antidotes, that these antidotes may in turn ultimately serve to transform life into *ars magna*, a great art.

I could not possibly have achieved this work alone. The spirit of Hermes has flowed through many kind souls who have helped me in ways innumerable and without whom this book may never have come into being. Even between the time of my great dream, which led me to become an astrologer, and the arrival of this volume, I have received too many kindnesses to list, and yet I know in my heart that each and every one of these truly matters. To all those who have lent a hand, a kind word or conversation, a drink, a meal, a ride, a job, or a place to stay—I honor you now, and pray for your aid through the spirit of Hermes.

There have also been some souls whose personal aid has been indispensable to this work, and I would like to thank them by name. Deepest thanks and honors to my sister, Irene, for all the basement lessons and cross-town walks to the library that awakened the spirit of Hermes within me, and to my wife Caroline, whose love and support have sustained me, my education, and my work for the last eighteen years. I am eternally grateful to the generosity of spirit, mind, and deed received from Robert Schmidt and Ellen Black at Project Hindsight. Thanks also to Michael Erlewine, who helped me renew and re-invigorate my classical studies there. Daniel Giamario and his work have been immensely important in reviving the ancient visual and mythic traditions of Mesopotamia. I thank him for that and also for his part in bringing out the ancient wisdom within my most important friend in this work, Adam Gainsburg. Adam has been like a brother to me, and without his generous sharing, encouragement, and wise example I could not have found my way to where I am now. Another big brother in this work has been Jeff Close, whose friendliness, generosity, Self-Evident astrology theory, computer-programming skills, and knowledge were a driving force long before I knew why Hermes had sent him into my life. It is through Jeff's invention of the Intrepid astrology program and his collaboration with Adam on the Sky Engine module that the path-breaking ephemerides in the appendix of this book are possible. And it was only through the deeper study of the perfectly formatted data these two brilliant astro-brothers put at my fingertips that the insights into the Emerald Tablet and deeper theories of this book were able to emerge.

Special thanks go to my good friend and alchemy tutor, Jim Rodgers, whose gift of a copy of the *Emerald Tablet* proved an invaluable channel for Hermes. Special thanks also to Michael Lutin for making me choose another topic and appreciating the promise in the Mercury material I came up with. Deep gratitude goes to Demetra George, for being a loving friend, resource, mentor, and colleague during the process of discovery and writing of this book. For the works of Erin Sullivan, Robert Blaschke, and Arielle Guttman along similar lines I am very thankful, especially for their personal encouragement which helped give me the audacity to attempt to stand on their shoulders and make my own contribution to the collective. And finally I am most thankful to Hermes for introducing me to my editors: Jenn Zahrt, for her enthusiasm for and belief in the value of this work as well as her patient and wise guidance, and Aaron Cheak, whose subtle Hermetic presence helped deepen this work in important ways.

Gary P Caton
June 6, 2016

with Mercury at maximum elongation as Morning Star, at 23° Taurus, conjunct Capulus—a star cluster in the sword hand of the hero, Perseus (he blessed by Hermes).

Introduction: A User's Guide

> *"Mercury might be the single most important planet...since it is through Mercury that we not only absorb but disseminate all of our perceptions."*
> —ERIN SULLIVAN[2]

WE LIVE IN a digital age of information. Successful symbol and image making, as well as information processing and delivery, are crucial in today's rapid-paced technological world. A broader, deeper, and more integrated understanding of how the archetypes of the planet Mercury function in our lives is therefore of utmost necessity. Hermes was once highly revered as the divine revealer of sacred mysteries, including writing, magic, and astrology. Beyond the merely practical then, a study of this archetype can bring one closer to the personal daimon, genius, or guardian angel that was classically seen to accompany each person from birth, and whose role it is to mediate between the divine and mortal. This volume is therefore ultimately fashioned as a catalyst for re-kindling the divine Hermetic awareness that served both as foundation for, and forerunner to, the scientific, industrial, and information revolutions.

Astrology has a long and complex spiritual heritage. In the religious and philosophical tradition of Hermeticism astrology is joined by the sister arts of alchemy and theurgy or magic. Together, these were once known as the three parts of the wisdom of the whole universe. It seems vital then to re-present astrology within the context of its original *hierophany,* or sacred revelation.[3] By examining all three of the intimately related Hermetic disciplines, we may form a more integral awareness, which is able to overcome some of the blindness and misunderstandings that have resulted from practicing astrology in isolation. In particular, by acquainting the reader with a more holistic and integrated view of the Mercury retrograde process, it is hoped this volume may serve to reframe Mercury's backward trickster medicine dance into something meaningful, transformational, and sacred once again. This sacred relationship with the god of transitions has the power not just to heal the individual and their life, but also to illuminate their relationship with humanity itself, and the basic human question they have incarnated to address and answer, for the good of the collective tribe.

Like the 125 years of triple alignments cataloged in the ephemerides found in the appendix, I have designed this compendium to be a lifetime companion and to deliver its rewards by multiple, diverse, and repeated readings. While I am sure an initial front to back read has its own unique rewards, it is far from the only way to use this volume. Periodically returning, and taking time to absorb and practice with the material in each chapter is both natural and advisable.

As Mercury's dance with the sun traces a great six-pointed star in the sky, with one triangle of this star symbolizing the three retrograde periods of each year, this book is divided into three sections and six chapters. These provide different windows into the same essential mystery. Each view is distinctive and yet contains enough of the whole to draw parallels and connections between the parts. This approach is known as *circumambulation*. By moving around the subject and exploring it from multiple points of view, the goal is to promote deep understanding rather than mere surface explanation. Another benefit of this approach is that, as we gradually explore cycles of greater and greater length, the reader also encounters the material in steps or stages. Hence, it is hoped this book can become many different books, even to the same soul, over the course of time.

SECTION I
The Mercury Elemental Year in the Birth Chart

This section introduces the reader to the astronomical phenomenon of Mercury's unique dance through the zodiac over the course of a year, and then applies this phenomenological awareness of the visual image of Mercury's motions to delineations of the individual chart or horoscope.

CHAPTER I: The Astronomy of the Mercury Elemental Year

For the last two thousand years the practice of astrology has largely been focused through the lens of the horoscope or chart of an individual moment, such as the birth of an individual person, country, or corporate entity. This chapter expands upon that now-conventional space and beckons one to enter a wider and more ancient territory and rekindle the awareness of astrology from the image-processing hemisphere of the brain. I have often found that in a consultation, when I simply explain the process of the astronomy like an illustrated story, many people spontaneously, and quite on their own, understand how the image formed by the astronomical phenomenon relates to their life.

CHAPTER II
The Mercury Elemental Year: Towards an Interpretive Paradigm

This chapter applies the visceral knowledge of Mercury's dance as a visual phenomenon to an individual natal chart or horoscope. The Mercury elemental year can activate the houses of a birth chart in four basic ways, so this chapter begins by identifying and delineating four basic groups of houses, or trigons. Each of these trigons is made up of three interconnected houses that define a set of related areas of experience that are heightened, with the particular areas depending upon the ascendant or rising sign of the individual. The result is 144 basic combinations, with examples included to allow one to begin the process of understanding possibilities inherent in each combination.

Section II
The Mercury Elemental Year as a Transformative Process

This section describes how to use Mercury's retrograde transits and their cycles through the four elements for personal growth. We explore various levels of spiritual awareness that can be activated by these transits, with information, analysis and exercises to re-sanctify and optimize the personal experience of Mercury's retrogrades.

CHAPTER III: Animism, Magic, and Myth

Animism as a worldview belongs with many of the earliest forms of human culture. Far from merely "primitive" ways of being which we have long since outgrown, a return or "descent" into more primordial ways of seeing and imagining can actually serve to re-vivify and en-soul our lives and world. This chapter helps us to understand the deeper story that Mercury tells us by re-examining the visual image of his sky journey. We thus gain the ability to use Mercury's visual descent as an opportunity to return to the source of human creativity and unleash our innate image- and story-making capacities.

CHAPTER IV: The Alchemy of Transformation

During the Middle Ages and Renaissance, astrology was often practiced as one of three "sister arts," which included alchemy. By reviewing some of the principles and philosophy of alchemy, we can better understand the process of transformation represented by the visual cycle of Mercury and its triple alignments. Once we understand the basic sequence of steps in this roughly forty-day transformational dance and the energies involved with each phase, we will then recognize its fractal nature and know how to dance with the rhythm of any transformational experience, at any scale.

CHAPTER V: Elemental Magic

The four elements actually pre-date astrology and Hermeticism, and they form the bedrock of many philosophical and esoteric traditions. Mercury's three sets of triple alignments usually occur in the same element each year. This means each of us was born with an unconscious preference for one magical mode of symbol-making and information-processing according to that element. Yet, after we are born the triple alignments continue to move through all four elements in a six- to seven-year cycle. By learning to follow this cycle and "making the quadrangle round," we can learn to practice the art of theurgy, integrating our inner opposite and becoming more and more whole.

CHAPTER VI: The Twelve Elemental Returns

This chapter lays the groundwork to understand the larger cycles of Mercury's returns to the element (and more rarely, the degrees) of one's birth (or of a particular endeavor). We thus gain insight into the cycle of life itself and its various seasons of becoming. Mercury returns to the same element every six to seven years, for a total of twelve returns within a span of seventy-nine years. Each of these returns correlates with a specific phase of human development, such that we can learn to follow certain quintessential trends as we journey through our lives and observe or advise on the lives of others.

SECTION III
Appendix: Mercury Elemental Year Ephemerides

This is the raw material from which the individual alchemist may derive the wisdom of Mercury and thereby work their magic. 125 years of Mercury's triple alignments with the sun are catalogued. These are conveniently arranged according to both the elements and degrees of the zodiacal signs. For the more advanced practitioner, the constellations and stars activated are noted as well. Together, this unique collection of raw data provides a rich astrological, mythical, and magical context within which to explore the elemental vintage, images, and meanings for any birth, event, or activity under consideration.

SECTION I
The Mercury Elemental Year in the Birth Chart

Chapter I
The Astronomy of the Mercury Elemental Year

> *"A beautiful sight, today seen by relatively few, Mercury shines brightly in the twilight sky...people of most ancient cultures were more familiar with the twilight wanderer and guardian of the sun."*[4]

IN THIS CHAPTER we discuss the transformative power contained in the image of Mercury's visual cycle. The dance of Mercury with the sun can be imagined as a circular ballet that is performed in three acts or movements. All of the basic steps of the dance are performed straight through, the whole set of dance steps is repeated, and then repeated once more. In this way, the dancers gradually make their way completely around the circular stage of the zodiac once, while repeating a set of movements about each other three times.

In astronomical terms, the relationship between Mercury and the sun is called the synodic cycle of Mercury. From our earth-bound (geocentric) perspective, Mercury completes a full set of movements about the sun every 116 days (on average). Thus, all given relationships between the two bodies are repeated three times within a 365-day calendar year. This means that if we plot the points at which a given relationship becomes exact, the three repetitions of this exact relationship will together form an approximate triangle within the circle of the zodiac.

Let us take for instance the "embraces" between the sun and Mercury. In astronomical and astrological terms, the coming together of two planets in the same space or degrees is called a conjunction. For Mercury and the sun there are two different kinds of conjunctions. The first kind is called the superior conjunction (from latin *superius*—meaning "above"), and occurs with Mercury in direct motion through the zodiac. This conjunction happens while Mercury and the sun are traveling in the same direction, with Mercury slowly catching up to, passing, and then moving ahead of the sun. The second kind of conjunction is called the inferior conjunction (from latin *inferius*—meaning "below"), and happens while Mercury is in retrograde motion through the zodiac. At this conjunction, Mercury and the sun are traveling in opposite directions, so that Mercury passes by the sun much more rapidly.

Plotting these two kinds of conjunctions for a given year, we see that each kind of conjunction forms a triangle with the two others of its kind. The points of these triangles activate the three signs belonging to a single astrological element. And together, these two triangles approximate a six-pointed star.

In astrology these triangles are associated with what are called the triplicities, another name for the three signs of the zodiac that belong to a single element. There are of course four triplicities or elements: fire, water, air, and earth. For the year in which I began writing this book, the superior or direct conjunctions were forming a triangle in the fire signs: Aries, Leo, and Sagittarius. Meanwhile the inferior or retrograde conjunctions were forming a triangle in the air signs of Gemini, Libra, and Aquarius. The triangle has always been a universal symbol of stability and harmony. Together, these interlocking triangles of Mercury suggest unity, wholeness, and integration between opposites.

Visually speaking, these conjunctions are completely invisible to the naked eye. Because of the sun's extreme brightness, no planet is visible when anywhere near the sun in the sky (that is, within approximately 15 degrees). Still, we have known about these conjunctions for a very long time because long before the Common Era, the Babylonian astronomer-priests learned how to extrapolate planetary positions using advanced geometrical and mathematical models to predict the future placements of the planets from their centuries of observations. Nevertheless, it is important to remember that to the naked eye the conjunctions are invisible because Mercury is far too close to the sun to be seen. As we move on, we will see that this hidden nature of the conjunctions represents an aspect of a "dance within the dance," that is, between the hidden and revealed sides of Mercury.

Another major feature of this dance is the two players appear tethered by an invisible and elastic cord. So, despite complete spins around each other and repeated comings and goings between embraces, the partners are actually never very far apart. The effect of this proximity is such that, throughout their gradual circling of the zodiacal stage, at no point in the dance will you lose track of one partner while focusing on the other.

Put in astronomical terms, from point of view of the earth, the sun and Mercury seem to appear as something of a binary pair. When mapping the sun and Mercury onto the circular zodiac (comprised of 12 signs of 30 degrees each, with a total of 360 degrees), we will never find them more than 28 degrees apart. Thus, despite a constantly changing relationship to the sun during their mutual movements through the zodiac, Mercury nevertheless always dwells in the same or adjacent sign to the sun.

Between invisible embraces or conjunctions, the dancers gradually move apart and then come back together, move apart and come back together, their invisible cord stretching and relaxing. Astronomically, the points between conjunctions of Mercury and the sun, where they are furthest apart, are called "greatest elongation." The amount of this elongation varies from cycle to cycle, but their maximum elongation for any particular cycle is never less than 17 degrees and never more than 28 degrees.

Just as there are two kinds of conjunctions between Mercury and the sun, there are also two kinds of elongations. Following superior conjunction, Mercury moves out ahead of the sun in the zodiac, so that at this maximum elongation Mercury is in a degree further along in the zodiac than the sun. However, following inferior conjunction, Mercury moves behind the sun in the zodiac. So, at this other maximum elongation Mercury is in a degree prior to the sun in the zodiac. In this sense, Mercury is constantly in the process of "switching sides" with the sun, moving ahead and then falling back.

Visually, the elongations are very different from the conjunctions, as it is the elongations that most reveal Mercury to the careful observer. In terms of naked-eye visibility, the best times to actually see Mercury in the sky are at maximum elongation. However, even at elongation, during the daytime, when the sun is in the sky, we still cannot see Mercury (or any of planets or stars) because of the sun's extreme brightness. Therefore, it is not only the times of Mercury's greatest elongations, but also only in the brief twilight times—the hour or so just before sunrise or just after sunset, when Mercury can best be seen.

In order to more easily differentiate between morning and evening elongation in an astrological chart, we first need to differentiate between the two different kinds of celestial motion. The zodiac is designed to track the yearly motion of the sun. As modern people we of course know the sun's *apparent* motion through the zodiac is actually a result of the earth's annual *revolution* around the sun.

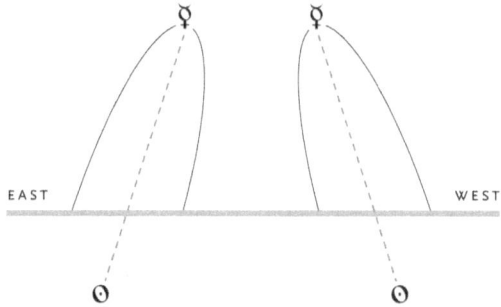

The figure on the left shows the maximum morning elongation of Mercury looking east. The figure on the right shows the maximum evening elongation of Mercury looking west. The solid elliptical line is Mercury's orbit, and the dashed line is the ecliptic, or the path of the sun.

The earth also has another kind of motion, which is its diurnal or daily *rotation* on its own axis, every 24 hours. These two kinds of motion are contrary to one another, such that in an astrological chart the apparent yearly revolution of the sun through the zodiac proceeds counter-clockwise whilst the apparent daily rotation of the sun from rising to culminating and setting proceeds in clockwise motion.

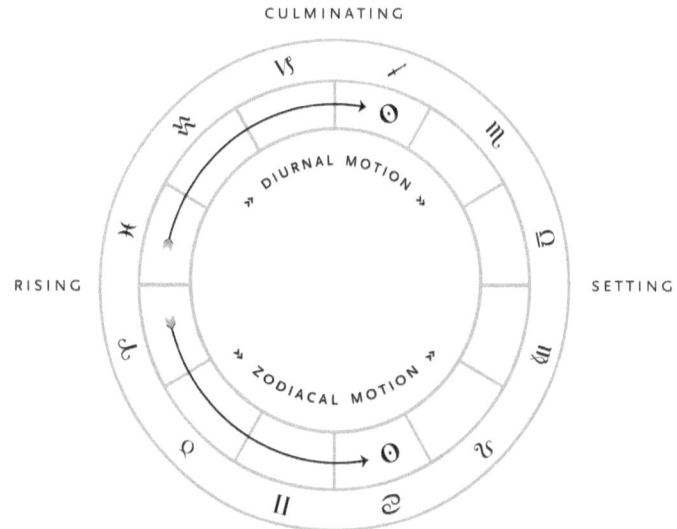

So, when Mercury is ahead of the sun in the zodiac it actually follows the sun across the sky in terms of the daily or diurnal rotation, rising and setting each day *after* the sun does. At this elongation, when the sun has already risen and it is daylight, Mercury cannot be seen in the morning sky, because the sun is too bright. It is only after the sun has set in the evening that we can see Mercury when it is ahead of the sun in the zodiac. Thus, we call this the evening elongation.

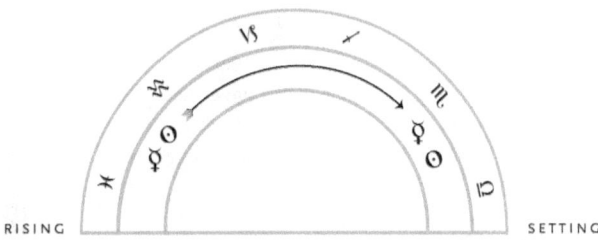

Conversely, when Mercury is behind the sun in the zodiac it actually precedes the sun across the sky in terms of the daily or diurnal rotation, rising and setting each day *before* the sun does. At this elongation, just before the sun rises, Mercury can already be seen above the horizon in the morning sky. Thus, we call this the morning elongation.

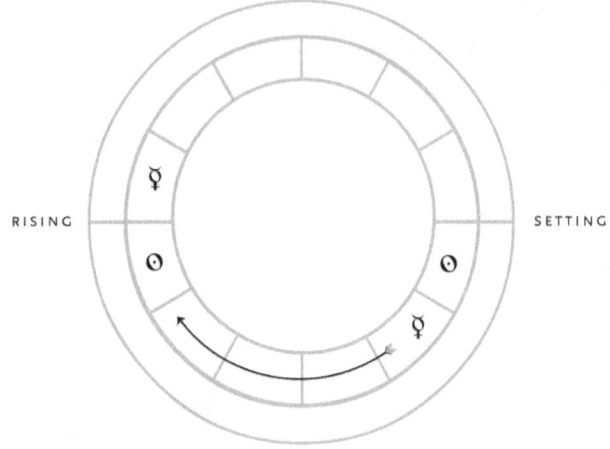

Now we have four key events in the synodic cycle of Mercury: two kinds of conjunctions and, between these conjunctions, two kinds of elongations. In order then, beginning at a superior conjunction, Mercury proceeds out ahead of the sun in the zodiac, stretching the invisible cord between them until reaching evening elongation. Mercury then begins to slow and the sun starts to catch up. Then Mercury appears to stop altogether, reverse direction (retrograde), and join with the sun at inferior conjunction. After this, Mercury begins to precede the sun in the zodiac. Upon resuming direct motion, Mercury is still slow, and the sun stretches out the distance between them until Mercury reaches morning elongation. After morning elongation Mercury begins to gain speed and gradually manages to catch the sun again, eventually passing the sun at superior conjunction. This entire back and forth dance then repeats itself twice more, for a total of three complete synodic cycles during a single calendar year.

At this point it might seem that we need to add something to the six-pointed star formed by Mercury's conjunctions in order to represent the elongations; however, that is not actually the case. Out of the four main events of superior conjunction, evening elongation, inferior conjunction, and morning elongation—significantly, the latter three all happen to occur within the same degrees of the zodiac. This is because the points of the triangle formed by Mercury's superior conjunction are passed only once, but the points of the triangle formed by Mercury's inferior conjunction are passed thrice! It is herein that the essential magic of the mercury elemental year has quite literally been "hiding in plain sight" within these triple alignments.

Although it is well known that Mercury takes about a year to traverse all twelve signs of the zodiac, what is less noticed is that he does not spend an equal amount of time in each sign during any year. The messenger planet shows decided preferences, often spending as much time in one elemental triplicity as the other three combined.[5] This extended stay in the signs of one triplicity is explained graphically in the triangle formed by Mercury's inferior conjunctions, whose points are also activated by the two elongations. This triple emphasis is very telling, and yet these triple alignments appear very different visually. They form what can be called a differentiated unity.

> "That which is Below corresponds to that which is Above, and that which is Above corresponds to that which is Below, to accomplish the miracles of the One thing."
>
> —HERMES TRISMEGISTUS,
> in *The Emerald Tablet*[6]

"As above, so below" is the simplified form by which this key passage is often evoked. The passage is often taken to mean that astrologers believe the causes of earthly events are to be found in the heavens. However, this is not quite accurate. When the entire passage is analyzed, the correspondence actually goes both ways: as below, so above *and* as above, so below. This implies an interaction, a process—an act of *co-creation*. In this light, the word "correspondence" can mean both a close similarity as well as an exchange of letters. I was most astounded to discover this similarity and exchange is not only spiritual or metaphorical, but also manifest—that is, literal, physical, and astronomical!

We know that three out of four of the most significant turning points in Mercury's dance with the sun, both elongations and the inferior conjunction, occur near the same zodiacal degrees. Visually speaking, we also know that the two elongations are visible above the horizon, while the inferior conjunction happens below the threshold of visibility. So, in terms of actual appearance to our eyes, this simply means that between elongations Mercury becomes invisible. As a visible phenomenon, Mercury retrograde is essentially a disappearing act. Mercury appears high in the west at dusk (evening elongation), gradually disappears below the horizon to join with the sun (inferior conjunction), and then re-appears high in the east at dawn (morning elongation). Mercury's retrograde motion is therefore synonymous with a journey into hidden or concealed realms.

The diagram below illustrates the synodic cycle of the planet Mercury as seen from the earth. This cycle repeats about every four months (116 days on average), or three times a year. Mercury makes his greatest elongation *above* the horizon in the western evening sky (1) before turning retrograde, disappearing *below* the horizon and making the "inferior" conjunction with the sun (2). Mercury then appears in the eastern morning sky, and again reaches his furthest distance from the sun, and so his highest appearance *above* the horizon (3). Significantly, *all three events always happen near the same zodiacal degrees,*—above, below, and above are indeed all *one*! Herein lies the deep profound magic of Mercury's dance.

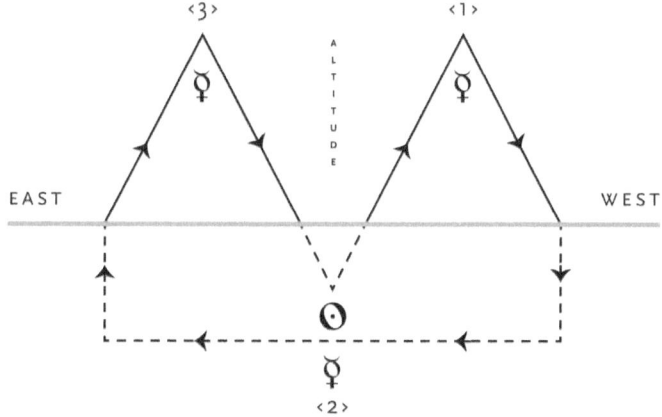

The sky view of Mercury retrograde: <1> maximum evening elongation; <2> Mercury retrograde conjunct the sun; <3> maximum morning elongation. The other sharp dip below the horizon without a number between <3> and <1> is Mercury's superior conjunction with the sun, which form the other complimentary triangle in Mercury's 6 pointed star.

Chapter II
Delineating the Mercury Elemental Year

*"Talent is like the marksman who hits a target
which others cannot reach;
genius is like the marksman who hits a target, as
far as which others cannot even see."*

—ARTHUR SCHOPENHAUER[7]

THUS FAR WE have seen how the three inferior conjunctions of Mercury with the sun each year form a triangle in the zodiac, whose points are crossed three times. We have also established that during retrograde motion Mercury is involved not only in a backward dance, but one that also finds him flitting in and out of visibility, moving from above the horizon to below, and then back above again. So, the power of this dance comes not just from the horizontal back and forth, but also from the vertical above/below part of the dance, which is just as important.

    Traditionally, these realms of above and below are associated with the principles "spirit" and "soul" in the alchemical literature. We will cover these terms and concepts more deeply in chapter four, but for now let us simply call spirit that which lifts a person up, and soul that which pulls a person deeper into life. Even with the tension between direct and retrograde—the status quo or establishment and the alternative or counter-culture—a horizontal life, lived primarily on the surface of things, is still largely bereft of both spirit and soul. We need the heights of thought, imagination, and aspiration and the depths of both love and pain in order to truly be alive. The transformative power of Mercury's retrograde derives from its juxtaposition and mixing of all four possibilities of direct/retrograde and above/below within the relatively short time span of around forty days.

All this rapid, repetitive, back-and-forth mixing between both direct/retrograde and above/below can be seen to correspond with significant eddies in the currents of collective consciousness. These eddies represent trends in thought, opinion, and events which not only run counter to the main current, but also link the above and below, the rational and irrational, the conscious and unconscious minds. The churning and stirring of these eddies therefore creates an interface between the establishment and the more progressive social factions, as well as between the spiritual and soulful parts of consciousness. The gifts of mixing these currents are the same gifts of diversity in a culture: it allows and creates new options and points of view previously unimagined, thus preventing the mind (both collective and singular or personal) from becoming overly fixated, stratified, stagnant, or sealed off.

Through this magical, mercurial triangle, in any given year the signs of one elemental triplicity will serve as the home of these counter-cultural and counter-conscious eddies, deeply informing the ways in which these eddies are able to bring up unconscious contents to the surface in order for them to be examined, transformed, and ultimately integrated. Just as the personal genius

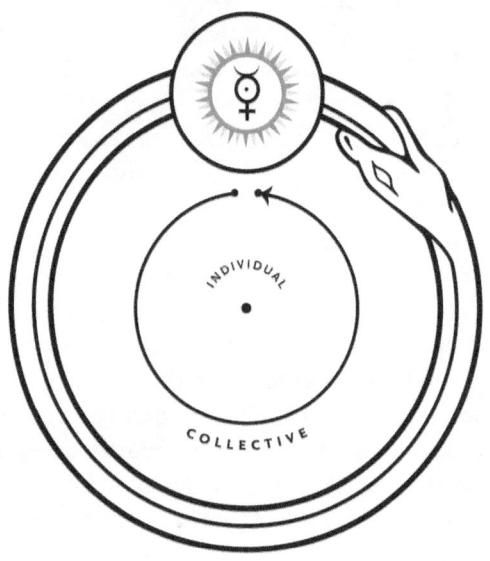

serves as mediator between the divine and mortal, the Mercury elemental year provides a collective context shared by everyone born in that particular year, which mediates between the singular/personal and the universal/collective unconscious. We can visualize this concept by using the glyph for the sun (see figure on opposite page). The dot in the center represents the individual chart or consciousness, but the circle around the dot represents a shared collective context—in this case the three signs of the Mercury elemental year.

Even if one is not born, or taking a particular action, during a time of Mercury retrograde—that birth or action still represents only a singular event which exists within a larger frame of reference. The eddies of the Mercury elemental year still form a contextual vessel, surrounding that event with counter-cultural and counter-conscious waves which serve to limit, shape, and define its collective context. As individual people, we are here to do just that—individuate. But our individuation also serves to answer basic collective human needs or questions, symbolized by the Mercury elemental year.

For instance, consider the basic human need for freedom and movement and the dream of taking flight, like the birds of the air. The first powered flight was made by the Wright brothers in 1903, a year in which the Mercury retrogrades were quite appropriately occurring in air signs. However, air is a volatile element, and accordingly the 1903 flights were short and unstable and so considered unremarkable by some at the time. The most remarkable accomplishments of the Wright brothers came later with their third Flyer and the completion of the three-axis aircraft *controls* that made extended fixed-wing powered flight possible and remains standard on fixed-wing aircraft of all kinds to this day. This three-axis system was perfected through trial and error by the brothers in 1905, a year in which the Mercury retrogrades were happening in earth signs, the most fixed and stable of the four elements.

Through the universal themes of the Mercury elemental year, there is correspondence between both the above and below, and the personal and collective. The collective currents of the Mercury elemental year serve to contain, shape, and define the individual and their efforts; at the same time, the individual also informs and ultimately serves as a vehicle through which these collective tides may be expressed. There is a correspondence and exchange made possible between the two. The three signs of the Mercury elemental year and the universal human needs they represent are in the process of being transformed both by and for everyone. A truly successful harnessing of these collective tides by any individual results in lasting changes to the collective possibilities for everyone.

Mercury is not commonly thought of in terms of collective expression, but since the Mercury elemental year is a context shared by everyone born in any particular year, it is precisely this collective expression of Mercurial themes to which the Mercury elemental year most clearly speaks. Each of us is born at a particular moment, which is entirely individual, but that moment is nonetheless contained within a collective context. Everyone born in the same year will have been born with the same three signs of one elemental triplicity involved in the mixing of conscious and unconscious currents.

In this way, the Mercury elemental year can be seen to illuminate a deeper field and context within which any birth or magical moment occurs. Just as all wines bottled by a vintner in a particular calendar year share the same general characteristics, similarly, the Mercury elemental year defines the vintage of your image making, information processing and spiritual transformation faculties, a vintage which you share with a good many other souls. Each particular vintage serves not only as a context or vessel for the individual but as a mandate for the individuals to lend their unique personal answers to the universal themes, needs and questions of that element. This is the target, heretofore unseen, toward which we now begin to ask our *genii* to take aim. To begin to see and understand this mandate, and the universal or collective human themes, needs, and questions we are here to address as individuals, we must find ways to frame this collective factor of the Mercury elemental year within a more personal context.

One way we can use this Mercurial triangle to see how these collective currents have an impact on our view of an individual chart or horoscope is through the places or houses, which are based on the individual factor of the ascendant or rising sign. The rising sign becomes the first place or house, and each sign proceeding through the zodiac takes the next number until the circle is complete. When viewed through this lens of signs serving as places or houses, the three signs of the element in which Mercury was retrograde during the year of one's birth signify specific areas of life where that individual is personally called (albeit often somewhat unconsciously, at least at first) to re-imagine, revise, and renew the consensus reality concerning the element of those signs, and the universal or collective human themes, needs, and questions which that element symbolizes.

More specifically, there are two signs in which Mercury most deeply informs the places or houses of any nativity: the signs and places or houses of the Mercury retrograde which occurred just before (pre-natal) and just after (post-natal) an individual birth both represent areas of life where the person will feel instinctively called to make a unique contribution to the collective—first, by breaking apart the old stagnant thought-forms of the status quo, and then by finding and bringing new contents welling up from the collective unconscious into personal expression to replace the remnants of a fading status quo. The use of pre-natal conjunctions, particularly lunations and eclipses, has deep roots going back at least to the Hellenistic era.[8] I have taken these inspirations a step further and included the post-natal conjunction as a way to create a container or contextual vessel for the birth or event, which occurs between, and is book-ended by, these two conjunctions.

In terms of the personal horoscope or birth chart, then, these two eddies connect a particular pair of houses or places within what is called a trigon, or group of three interrelated houses. The house containing the pre-natal inferior conjunction of Mercury and the sun will be one in which the individual feels restless and instinctively drawn toward unconventional expressions that run counter to established traditions. They are here to show us how the answers we have come to accept for these basic themes, needs and questions have grown stale or inapplicable. The personal expression of these collective eddies thus serves to break apart the established order, which in turn allows space for something new to emerge. The house containing the post-natal inferior conjunction of Mercury with the sun will be one in which the individual feels instinctively called toward communicating or expressing in an avant-garde fashion, seeking out new and cutting edge expressions that are arising from the collective unconscious. The personal expression of these counter-cultural and counter-conscious eddies serves to bring together and establish a new order—like a seedling struggling to emerge from the universal ground of the collective unconscious. These new answers to the timeless collective human themes, needs, and questions serve to keep our societies alive and growing, capable of responding to change.

Though the houses of the pre-natal and post-natal conjunctions are most important, we must remember that all three of these houses of the trigon are connected and pregnant with new possibilities due to the mixing of the direct and retrograde, above and below. Still, because these currents are bubbling up alternatives from the collective unconscious so close to birth, a person is especially likely to feel a push away from the traditions of the house containing the pre-natal conjunction and a pull toward new modes of expression in the house containing the post-natal conjunction. Since the personal chart is usually most immediately and deeply affected by the collective currents from the two signs and houses of the pre- and post-natal inferior conjunctions, the following delineations will primarily focus on the synergy between these two areas. The house of the pre-natal conjunction will show the primary area in which the native unconsciously reacts against and instinctively runs counter to the cultural norms of the status quo, and the post-natal conjunction will show the primary area where the native can see and feel newly emergent forms of the collective unconscious better than most people.

Even while being conscious that two signs and houses are most significant for any individual, we must still delineate these two signs and houses through a lens of wholeness, one which includes all three signs of the triplicity activated in the birth year. This is what astrologers call a trine relationship, and it reflects a harmonious relationship that allows for an easy flow of energy. The three signs of an elemental triplicity are intimately related, flowing together and affecting one another like the members of a family or a group of connected bodies of water. These three signs act as a system, so that even small changes in any one of these areas will eventually have ripple effects throughout all three, and because all three areas are so deeply interconnected, we must maintain awareness of all three houses or places in any individual example.

In this way, the triplicity of the Mercury elemental year gives us four different trigons of houses that can be activated within an individual chart. Thus the collective currents of the Mercury elemental year have four fundamental ways to play out on an individual level. These four trigons consist of the following sets of houses: 1/5/9; 2/6/10; 3/7/11; and 4/8/12. Because angular houses have always been seen as more powerful by initiating important and essential activity, the angular house amongst any of these four trigons defines the primary purpose of the entire trigon. Following this logic, I call houses 1/5/9 the Identity Trigon, houses 2/6/10 the Trigon of Mastery, houses 3/7/11 the Relationship Trigon and houses 4/8/12 the Trigon of Dynasty.

While the angular house is the most important of each trigon—because it initiates the essential activity defining that trigon's purpose—the cycle of energy exchange can only be completed via the succedent and cadent houses, so they are just as vital to the process. After the angular house initiates activity, the succedent house then serves to stabilize the trigons purpose, by solidifying and holding the fruits of those activities. Finally, the cadent house serves to distribute the fruits of the trigon, which in turn naturally catalyzes further action (back in the angular house). This natural cycle of perpetual energy exchange is what drives the processes of the Mercury elemental year.

When delineating the Mercury elemental year in a personal horoscope, we start by locating the signs of the inferior conjunctions that occurred before (pre-natal) and after (post-natal) the birth date (or beginning of an enterprise). These signs may be found in the ephemerides in the appendix. Once these two signs are located, then determine their places or houses, by their relationship to the ascendant or rising sign of the chart (which is calculated from the date, exact time, and place of birth). Once the ascendant is calculated, the places or houses of the pre- and post-natal conjunctions can be ascertained simply by counting through the zodiac sign by sign, with the ascendant or rising sign being place or house number one, the next sign in zodiacal order becoming house number two, etc. For the vast majority of people (about eighty percent), both of the signs of the pre- and post-natal conjunctions will be located in a single element and trigon. As an example, if the personal chart has Aries rising and the signs of the pre- and post-natal inferior conjunctions are Leo and Sagittarius, then in the following delineations you will consult the section on the Identity Trigon, and the sub-section regarding houses five to nine.

The following delineations also contain notable examples to illustrate the possible embodiments of these combinations. These examples are taken from Astro-Databank records, which are available online. If you would like to see any of the charts of these celebrity examples, simply conduct an online search for the name of the celebrity and add the term "astrodatabank." Great care was taken to be as inclusive and diverse as possible in choosing these examples, in order to obtain a truly representative cross-section of gender, race, culture and other factors. It is hoped any perceived shortcomings by the reader in this regard will be understood as an unintentional by-product of the data itself, which is of course subject to various social biases. Regardless of these possible biases, the data can be judged to be very accurate in the technical sense, as it was painstakingly culled from about ten thousand records of public figures with data having a Rodden rating of A or better. This rating means the time of birth comes either from an official record itself or the record as quoted by the person, kin, friend, or associate. Thus is it very likely these birth times are accurate.

Beyond the scope of these delineations are examples from the roughly twenty percent of people who will be born at a time where the Mercury elemental year is in a state of flux, moving from one element to another. The transition between elements suggests these individuals are here to make important transitions, both in their personal growth and karma, as well as in their dharma or collective contribution. This can mean there will be a significant and important transition in the life of the individual, such that they live two very different lives within one lifetime. It may also means that they can learn to live their whole life with one foot each in two very different worlds, like some of the images in the major arcana of the tarot (e.g., both the Temperance and Star cards tend to contain figures with one foot on land or in the earth element and one foot in the water). This is a very transformative position and can be a more difficult situation than dealing with a single element, yet it can also potentially be much more powerful and rewarding, particularly when two angular houses are activated. These individuals will benefit by paying special attention to the material in chapter five, which explains the process of transformation between elements.

In terms of trigons, when the birth date (or beginning of an enterprise) is such that the previous inferior conjunction is in one element and the following inferior conjunction is in another element, then it will be necessary to consult two different sections in the delineations of this chapter. The pre-natal element/trigon is more likely to dominate consciousness early in life, urging the native to break apart from the normalcy of the status quo regarding the current expressions of that element and house. Later in life, beginning as early as adolescence and culminating after mid-life, the person will begin to feel the pull toward the post-natal element/trigon. They will feel drawn to channeling new forms of expression for this element from those just beginning to emerge from the collective unconscious into public awareness. The reader must reflect on the delineations from the two different trigons and develop an intuitive feeling for which of these two is currently manifesting or dominating the awareness and life. Ultimately these two trigons, houses and elements will need to be integrated. There are some examples of this process at the end of this chapter, after all the permutations of the individual trigons have been delineated.

Less than one percent of people will be born on the very day of an inferior conjunction. In these cases it is advisable to begin the delineation by treating the conjunction of the birth date as if it were the post-natal conjunction, at least up until early adulthood. At some point in their lives, and this will vary, these individuals will begin to be drawn toward the house of the post-natal conjunction, and at that point the inferior conjunction of the birth date can be read as the pre-natal conjunction. Thus it will be necessary for the astrologer to consider all three of the houses of the trigon activated for these individuals, and then use open-ended questions and intuition to ascertain where the client is currently located within the process of transformation represented by all three houses of the trigon.

THE IDENTITY TRIGON: Houses One, Five and Nine

Because angular houses have always been seen as more powerful by initiating important and essential activity, the angular house defines the primary purpose of the entire trigon. The angular house of the 1/5/9 trigon is the first house and so I call this the Identity Trigon. So, whereas the angular house initiates important and essential activity, the succedent house serves to stabilize this purpose by solidifying and holding the fruits of those activities, and the cadent house then serves to distribute these fruits, which in turn catalyzes further action. We can see then that the Identity Trigon is made of three basic parts: the personal identity and roles we take on in order to play a part and have a voice in social discourse (first house); the creative expressions and desire to take risks that naturally comes from having a voice or social agency (fifth house); and the higher viewpoints and cultural expansion and exchange that results from successful creative expression and risk taking (ninth house).

It can be construed that individual souls born at times of day such that the three Mercury retrogrades of their birth year are activating the three houses of the Identity Trigon are incarnating with a deep rooted, instinctive, and somewhat (at least at first) unconscious drive to re-frame the collective possibilities within and from which individuals may forge a personal identity. They are here to break apart the collective roles currently available for and assigned to individuals and eventually to put these roles back together, either into new creative combinations, or sometimes to help forge entirely new roles within new cultures or between existing cultures.

As an example, given the actual fluidity of gender, we can expect that a soul born with the Mercury retrogrades of their year activating the Identity Trigon may be somewhat desirous of expanding their personal expression beyond the confines of normative, binary gender associations. This is not necessarily a "queer signature" (though that may be one possible expression of it), but the person may simply tend toward more androgynous or nonbinary expressions.

Though they are not necessarily directly related, both sexual preference and gender identity exist on a continuum. In terms of gender identity a person in tune with the status quo will tend to be sex-typed, that is displaying mostly psychological traits culturally associated with their physical sex. However, we can expect that souls born with the Mercury elemental year activating the Identity Trigon may be more naturally androgynous, that is, displaying psychological traits culturally associated with both genders, or even on a spectrum outside of the gender binary. Thus a person assigned male at birth with the Identity Trigon activated could be drawn to expressing some aspect of the "lunar masculine," and a person assigned female at birth with the identity trigon activated may be drawn to expressing the "solar feminine." While it is important for astrologers to respect and honor a wide range of personal expressions of gender, it is equally important to remember that fluidity with regard to gender identity is but one possible manifestation of many with this trigon, even within the delineation of a single individual chart.

The person with the Identity Trigon activated may inhabit multiple roles, which can be seen to exist on the fringe of the status quo. A multiplicity of identities which conflict with elements of the dominant elements of society necessitates awareness of the concept of intersectional theory, which is the study of overlapping or intersecting social identities and the multiple interrelated systems of oppression, domination, or discrimination to which an individual may be subjected as a result of these identities. Many people identify with multiple statuses currently considered to be in the minority, and even a person who at first glance seems to enjoy the privilege of identifying with the dominant status quo may still be fighting multiple battles with regard to gender identity, religion/spirituality, social class, age, and/or living with disabilities. More than simply respecting and honoring diversity, which is a good start, astrologers are tasked with learning, acknowledging, and supporting the heroism it takes to inhabit all kinds of roles and help affirm and explore the extremely valuable collective contributions they make possible.

Delineating the pre-natal and post-natal inferior conjunctions by (whole sign) house:

Houses One to Five
When the pre-natal inferior conjunction of Mercury and the sun is located in the first house (and thus the post-natal inferior conjunction usually in the fifth) the person will somehow—often somewhat unconsciously, at least at first—embody a particular image, style, or role which runs counter to the collective norms of their society. Their free-spirited nature holds a magnetic appeal, and this, in turn, inspires both them and others around them to take new risks in their personal creative expressions (post-natal conjunction in the fifth). These people possess the potential ability to become the personal embodiment of newly emergent social identities, which

at once signify a break with the past and an invocation of a future with new freedoms of personal expression. These are authentic individuals, and once they become conscious of their unique contribution, they can serve to transform society by successfully inhabiting fringe roles to the point where these personae naturally become more and more accepted, adopted, and integrated into their social structures.

FIRE SIGNS: Aries, Leo, and Sagittarius

The fire persona in combination with the Mercury retrogrades also in fire signs makes the natural exuberance and volatility of the fire element especially noticeable with this placement. Fire is both active and separative and tends to move up and out, seeking the rarefied presence of pure spirit by casting aside or burning through gross materiality. For *Aries rising*, we can see these principles embodied by **Joan Rivers**, who used her acerbic wit and volatile, controversial comedic persona to break through the barriers of identity and became the first woman to host one of the late-night talk shows on major television networks. Rivers' unabashed embrace of plastic surgery as it was becoming more prevalent also exemplifies the influence of pre-natal Mercury retrograde in the first house. **Tina Turner** shows this fiery presence quite clearly as a *Leo rising*. Turner's choice of the volatile life of a rock star involved rising above the early experiences of an abusive relationship through the embrace of non-mainstream religion (Buddhism), and her through her subsequent solo career she became one of the most well-known and successful performers in the world. **Dame Elizabeth Taylor** (*Sagittarius rising*) began as a childhood star but had to fight through scandal and condemnation from both the Vatican and US Congress to achieve critical success. Taylor's most acclaimed performance came in *Who's Afraid of Virginia Woolf?*—the film was considered groundbreaking for its overt treatment of sexuality and provocative language. The film nonetheless garnered awards, and this helped lead to the formal ending of the "production code," used to determine what was morally acceptable in motion pictures, from 1930 to 1968. Taylor later helped pioneer celebrity support for AIDS activism in the mid '80s, long before it became fashionable.

AIR SIGNS: Gemini, Libra, and Aquarius
The air persona in combination with the Mercury retrogrades also in air signs highlights the natural sociability and communicative nature of the air element. Air is very active but also connective, and so a bit more stable than fire. Thus, once air moves up and out, it can stabilize and somewhat paradoxically prefer to see or experience things from a safe intellectual or social distance. For instance, the *Gemini rising* native **Mick Jagger** became the epitome of the bad-boy rock star with his early drug use and romantic involvements and as the lyricist for one of the most successful songwriting partnerships in history, with guitarist Keith Richards of the Rolling Stones. However this volatility mellowed, and after solidly establishing themselves, the Stones became one of the most enduring acts in Rock and Roll history. For *Libra rising*, we can see these principles demonstrated by **Britney Spears**, who parlayed her initial success in *The Mickey Mouse Club* into pop super-stardom as a teen in the late '90s. Her later excursions into reality television, relationship, and parental scandals received heavy criticism, and courts placed her under an ongoing conservatorship in 2008. Despite these enormous personal struggles, her unique combination of innocence with a dark side continues to result in iconic status and enduring success. We can see the airy presence of *Aquarius rising* notably demonstrated by film and fashion icon **Audrey Hepburn**, who played the poor waifish country girl Lula Mae Barnes in *Breakfast at Tiffany's* who transformed herself into Holly Golightly, an eccentric urban trickster in the New York café society. This performance is considered a defining moment for modern women.[9]

WATER SIGNS: Cancer, Scorpio, and Pisces
The water persona in combination with the Mercury retrogrades also in water signs highlights the natural introspective, imaginative, and caretaking qualities of the water signs. Water is connective like air, but water is denser than air, so water moves down and in, eventually conforming itself to the structure and boundaries it encounters, and this often results in a universal down-home feel. Since all the action is often below the calm or relatively uniform surface of water, a great diversity of talents and interests are notice-

able in those with this placement. For instance, *Cancer rising* native **Stephen King**'s characters tend to portray seemingly normal people who have hidden within them unordinary psychic powers or some other personal relationship with the supernatural. King's talent for writing everyday dialogue and revealing deep emotional and spiritual insights have made his work extremely popular. We can see the fixed watery presence of *Scorpio rising* in stock car racing pioneer **Ralph Earnhardt**, whose toughness, determination, and innovations transformed him from a poor cotton mill worker into the founder of a racing dynasty, with both his son Dale and his grandson Dale Jr. going on to become legendary NASCAR drivers. Multi-talented film star **George Clooney** exemplifies the universality of water via *Pisces rising*. Clooney began as an extra in a television series filmed in his hometown, maturing through a decade of work as a popular television actor. After transitioning to the "big screen" of the film industry, Clooney went on to directing and producing, becoming the first person to be nominated for an Oscar in six different categories.

EARTH SIGNS: Taurus, Virgo, and Capricorn
The earth persona in combination with the Mercury retrogrades also in earth signs results in the natural talents for developing technique, synthesizing a variety of influences, and cultivating classic, everyman appeal common to the earth element. Earth is the densest of elements and also the most solid and stable. Like water, the relatively uniform surface of earth often hides a surprising diversity, and the separative nature of earth is adept at revealing structure and strata. As part of the New Hollywood movement, *Taurus rising* native **George Lucas** exemplified the new forms of expression often natural to the identity trigon via *Star Wars*, one of the most outstanding and beloved motion pictures of all time, which synthesized many of his early visual, literary, and spiritual influences. Initially talented in camera and editing work, Lucas was introduced to screenwriting by Francis Ford Coppola and gained a sense of mystery, spirituality, and mythic structure from the work of Joseph Campbell, among others. Legendary Austrian mountaineer **Hermann Buhl** (*Virgo rising*) was the first climber to break the

8000-meter barrier while climbing solo and without bottled oxygen. Buhl is widely regarded as one of the best climbers ever and was particularly pioneering and innovative in applying the Alpine style to the Himalayan peaks, which has gradually become more the standard. The earth identity is exemplified in its cardinal form by *Capricorn rising* native **Anthony Hopkins**, who transformed himself from the son of a baker into one of the best actors in the world. Having success first in the theater, as well as a wide variety of roles in film, Hopkins' shocked the world as Dr. Hannibal Lecter in *The Silence of the Lambs*, a performance which broke ground as one of the shortest ever to win the Oscar for Best Actor.

Delineating the pre-natal and post-natal inferior conjunctions by (whole sign) house:

Houses Five to Nine

When the pre-natal conjunction is in the fifth house (and thus the post-natal conjunction usually in the ninth), there is an (unconscious) urge to explore creative options, take risks, push limits, and explore the boundaries of personal identity. This urge may at times seem unreasonable or irrational to others, and the individual may indeed eventually need to learn to modulate its expression. However, it should not be repressed or overly controlled, but rather encouraged and given means of safe expression. These individuals contain the potential to change the collective by taking risks that were not previously possible or by spending newfound creative capital in a way that opens up new frontiers. By pursuing their personal passions beyond prior reasonable limits, they can potentially expand the realm of possibilities available to us all (post-natal conjunction in the ninth).

FIRE SIGNS: Aries, Leo, and Sagittarius

The upward and outward striving, inherently passionate nature, and tendency to break down structures of the fire element can all be expanded with this position—sometimes to the point of dangerous or possibly even destructive intensity. However, at times it is precisely this willingness to risk or do it all that provides the fuel to move creative efforts beyond the personal sphere to change the culture as a whole. For instance, while perhaps best known today for her role in Theosophy, *Aries rising* native **Annie Besant** was also instrumental in politics, such as in the London Matchgirls' strike of 1888 and the successful London Dock Strike. Her passionate support of both Irish and Indian self-rule is a classic example of both the natural independence of fire and the natural anti-establishment transformative influence of the identity trigon moving from the personal to cultural (post-natal conjunction in the ninth). For *Leo rising*, we can see these principles demonstrated by **Galileo Galilei**, who in 1610 made extraordinarily detailed and pioneering observations that laid the foundation for modern physics and astronomy. By empirically confirming the Copernican heliocentric system, his knowledge was far ahead of its time, but because of the volatility this new knowledge represented to the prevailing religious order of the day—and the volatile way he presented it—he was eventually condemned to house arrest by the Inquisition for the remainder of his life. **Leonardo Da Vinci** (*Sagittarius rising*), still celebrated as one of the most talented people in history, employed empiricism before it became a standard operating procedure, and his artistic legacy matches his contributions in science and engineering.

AIR SIGNS: Gemini, Libra, and Aquarius
The natural predilection for upward movement and the social and intellectual awareness of the air signs can be expanded beyond the personal sphere with this position, with the ensuing cultural impacts having both inspirational and potentially worrisome possibilities and consequences. For *Gemini* rising, we can see these principles embodied by **Buzz Aldrin**, who became the second person ever to walk on the surface of the Moon. Aldrin later advocated further expansion of human space travel and presented a plan not just to land on but also to colonize Mars by 2040. *Libra rising* native **Antonin Scalia** pushed the boundaries of the identity trigon by becoming the first Italian-American justice on the US Supreme Court. Scalia's thirty-year term as justice is notable for his support of extremely controversial decisions, and the con-

servative majority he was part of massively expanded the potential amount of money allowed for spending in American political elections via decisions like *Citizens United* and *McCutcheon v FEC*. Surpassing traditional gender roles, **Irene Miller Beardsley** (*Aquarius rising*) was one of two women to summit Annapurna, the deadliest 8,000-meter peak in the world, bringing success for the 1978 American Women's Himalayan Expedition. As the first US team to summit the mountain, the women creatively anticipated later generations' more mainstream use of crowdfunding by raising three-quarters of the expedition's $80,000 costs through selling t-shirts with the slyly brilliant slogan: "A Woman's Place Is On Top: Annapurna."

WATER SIGNS: Cancer, Scorpio, and Pisces

The natural sensitivity, amorphousness, and universality of the water signs can be expanded beyond the personal sphere with this position, leading to visions of cultural or even cosmic unity, but with the danger of such unbounded expressions as cannot be practically understood or applied. For instance, *Cancer rising* native **Steven Spielberg** is considered one of the founding pioneers of the New Hollywood era, with his early films seen as archetypes of the modern Hollywood blockbuster. Spielberg expanded his creative focus from the merely shocking exposure of hidden fears in the thriller *Jaws* to explore more serious, cross-cultural, and humanitarian themes such as slavery, war, the Holocaust, and terrorism. For *Scorpio rising*, we can see the concept of identity expansion exemplified by **Gloria Steinem**, who first achieved national fame as a columnist for *New York* magazine, and then moved into publishing as a cofounder of *Ms. Magazine*, a popular quarterly periodical made for and by women. Steinem's activism also extends beyond feminism, with her focus directed to broader issues such as child abuse, the death penalty, race, and equality. *Pisces rising* native **R. D. Laing** expanded the focus of identity to include accepting the feelings of people struggling with apparent mental illness as valid experience. Laing rejected the purely medical model that was orthodox for his day and proposed instead that what is called mental illness results from the natural psychological tension between fundamentally different identities. For instance, all of us wrestle with contradictions between our social identity, defined by a desire to please our family and society, versus establishing our own unique and authentic personal identity. Even in "normal" people, the necessary compromises between identities results in a certain amount of alienation, which they learn to cope with and accept. What the majority label as abnormal or illness, Laing would deem "differently alienated," and creative people are more at risk of being so, because of their willingness to explore.

EARTH SIGNS: Taurus, Virgo, and Capricorn
The earth signs' natural tendency to explore diverse interests and talents through hard work and skillful practice can be expanded beyond the personal sphere with this position, sometimes leading toward a greater collective or cultural identity but often with the constant risk of fracture, re-division, or stratification. For instance, *Taurus rising* native **Mary Stuart** was claimant to the crowns of four nations but was eventually forced to abdicate her inherited identity as Queen of Scotland. The Catholic Mary met with suspicion from her Protestant cousin Queen Elizabeth I of England, and Elizabeth ultimately executed her for treason. However because Elizabeth was without an heir, Mary's son James VI eventually became the King of England (as James I), thus extending the collective identity of the Stewart dynasty from Scotland to all of Great Britain and inaugurating the Jacobean era, famous for its arts and literature, including Shakespeare and the King James Bible. The expansion of identity through developing diverse practical talents is exemplified by *Virgo rising* native **Warren Beatty**, who became only the second person to receive Oscar nominations for writing, producing, directing, and acting in the same film. Beatty later became the first person to achieve the feat twice, for two separate films. **George Steinbrenner** (*Capricorn rising*) helped push the boundaries of identity for major league baseball athletes through his embrace of the new era of "free agency." Steinbrenner's arrival as owner of the New York Yankees came at a time when players' union challenges overturned the exemption from anti-trust for baseball. Unafraid of committing to rising salaries, Steinbrenner aggressively pursued

the biggest stars and was the first to build winning teams under the new system.

Delineating the pre-natal and post-natal inferior conjunctions by (whole sign) house:

Houses Nine to One

When the pre-natal conjunction is in the ninth house (and thus the post-natal conjunction usually in the first), there is an (unconscious) urge to expand upon new cultural expressions and integrate them into a unique and diverse personal identity. By taking the next steps to both broaden and deepen the cultural frames of reference within which they express their personal passions, they can forge a new unity—forming a singular, idiosyncratic outlook out of many viewpoints. There may be a seemingly irrational or unreasonable urge to seek sources of information or inspiration beyond what the familial and societal milieus provide. This searching is not a rejection of the natural inheritance, but rather an attempt to re-fecundate and diversify to find combinations that represent the best next evolutionary step. There may also be a tendency for these individuals to want to be a catalyst regarding changing the guiding philosophy of their society such that the native becomes a spokesperson for newly emergent causes or perhaps even personally seeks to change the course of world affairs. These individuals instinctively gravitate toward roles where they can take on a larger than life appearance or persona and may even become an icon or symbol for the collective.

Fire signs: Aries, Leo, and Sagittarius

The natural optimism and questing nature of the fire signs can not only expand into a larger than life vision here, but potentially bring it back down to earth as well, and become the personification of a new collective identity. These individuals aim high and have the potential power to change the world, however the volatile pursuit of collective glory also carries the potential for burnout and unrealized dreams. For *Aries rising*, we can see these principles embodied by **John DeLorean**, who was a successful engineer and executive in the US automobile industry before going on to found his own DeLorean Motor Company. Though his company had a brief and turbulent history, DeLorean's innovative spirit lives on in the form of the iconic image of his DMC-12 sports car transformed into a time machine by an eccentric scientist in the 1985 film *Back to the Future*. Astrologer **Demetra George** (*Leo rising*) integrated the expanding feminine social identity of the '70s into astrology by helping to pioneer the use of asteroids in chart interpretations. George was later part of other pioneering ventures that sought to expand the collective identity of astrology, both in the restoration of Hellenistic era astrology theory and practice and as a teacher at Kepler College. We can see these principles embodied by *Sagittarius rising* **Angelo Giuseppe Roncalli,** who eventually became known as Pope Saint John XXIII and confounded expectations that he would be a "caretaker" pope. By calling the historic Second Vatican Council (1962–65), he helped change the face of Catholicism and its relationship to other religions and the modern world.

Air signs: Gemini, Libra, and Aquarius
The natural communicativeness and sociability of the air signs can expand into a larger-than-life vision with this placement. These individuals aim for high social goals and have the potential to change society, if not the world. However, the pursuit of collective glory also carries the potential for losing touch with one's roots. For instance, *Gemini rising* native **Henry Kissinger** fled Nazi persecution to become a US citizen and decorated war veteran, and he eventually had a long, influential, and controversial career in diplomacy, playing a prominent role in US foreign policy for almost a decade. For *Libra rising*, we can see these principles evinced by **Arthur Ashe**, who pushed the boundaries of identity by becoming the first black man to win grand slam titles in tennis. Ashe's win at Wimbledon is seen by many as particularly memorable. Ashe was the first black player on the US team for Davis Cup and controversially said his opponent Connors was unpatriotic for playing a lucrative exhibition match instead of representing his country. On top of this, Ashe had to change his game considerably to defeat the much younger Con-

nors, so the win is remarkable in several contexts, including personal, athletic, strategic, and social. Ashe went on to use his celebrity status to raise awareness about apartheid and AIDS. **Whoopi Goldberg** (*Aquarius rising*) demonstrates these principles by being one of the few entertainers to have won a "show business grand slam" or EGOT (Emmy, Grammy, Oscar, and Tony Award). Goldberg also used her celebrity status to expand identity awareness through high-profile support for LGBTQ+ rights and AIDS activism.

Water signs: Cancer, Scorpio, and Pisces

The natural sensitivity, depth, and universality of the water signs can expand into larger-than-life themes with this placement. These individuals aim for high social or spiritual goals. However, the boundless and shapeless qualities of water can become dangerous if carried too far. For instance, Nobel prize–winning mathematician **John Forbes Nash Jr.** (*Cancer rising*) changed how we understand chance and decision-making, despite struggling deeply with mental illness. After leaving life as a Catholic monk and later being denied tenure as a university professor, *Scorpio rising* native **Thomas Moore** went on to become a psychotherapist and bestselling writer who lectures internationally on deepening spirituality and cultivating soulfulness, even in the medical field. For *Pisces rising*, we can see these principles evinced by **Richard Pryor**, who expanded identity roles by elevating stand-up comedy to an art form and thus paving the way for further generations of comedians. Just as the water element conceals possible dangers in its depths, Pryor was the master of the uneasy laugh—a joke that is simultaneously hilarious and disturbing because it carries an edge by being built on a deeper, more disturbing social truth.

Earth Signs: Taurus, Virgo, and Capricorn

The natural diversity of practical talent and everyday appeal of the earth signs can expand into larger-than-life themes and success with this placement. These individuals naturally aim for lasting success and change. However, there is a danger of crumbling or fracturing if not deeply grounded in solid principles and structures. For instance, *Taurus rising* native **Toni Morrison** helped bring black literature into the mainstream first as an editor and later as a Pulitzer and Nobel Prize-winning author. Morrison completed her first novel as a single working mother of two by rising every morning at 4 AM to write. We can see these principles evinced by *Virgo rising* native **Dame Ellen MacArthur,** who once set a new world record time for circumnavigating the globe solo. After retiring from sailing, she began a charity to take people sailing as an aid to their recovery from cancer and a foundation to look beyond the current extractive throw-away society, towards a "circular economy" designed to mirror the regenerative properties of nature. During his rough youth in South Central Los Angeles, **Barry White** (*Capricorn rising*) decided to turn his life around after hearing Elvis Presley. White worked in singing groups and as a producer before achieving phenomenal success as a solo musical artist, becoming one of the best-selling artists of all time.

Note: In the Identity Trigon, the element of Mercury's retrogrades matches the element of the ascendant or rising sign. An important consideration in the delineations for the next three trigons is the elemental shift between the signs of Mercury's retrogrades and the element of the ascendant. The Trigons of Mastery and Dynasty contain difficult combinations between polarized elements that do not share qualities or naturally combine. The Relationship Trigon is more similar to the Identity Trigon, being made up of combinations of compatible elements with shared qualities that more naturally combine. These elemental shifts become more evident when drawn in a chart, but it bears repeating here. It will help to sketch out a rough chart (similar to the one on page 36) as the delineations progress, in order to see how the elements shift around the trigon and ascendant combinations.

THE TRIGON OF MASTERY: Houses Ten, Two, and Six

Because angular houses are more powerful by initiating important and essential activity, the angular house defines the primary purpose of the entire trigon. The angular house of the 10/2/6 trigon is the tenth house, and so I call this the Trigon of Mastery.

The angular house initiates important and essential activity; the succedent house serves to stabilize this purpose by solidifying and holding the fruits of those activities; and the cadent house then serves to distribute these fruits, which in turn catalyzes further action. We can see then that the Trigon of Mastery consists of three basic parts: the public roles (tenth house) which we take on in order to serve as representatives of the collective; the personal capital, both tangible and intangible, that we have naturally and that we acquire (second house); and the daily environment and routines (sixth house) which we draw upon in order to practice and refine these public roles.

Individuals born at times of day when the three Mercury retrogrades of their birth year activate these three houses enter this world with a deep, instinctive, and somewhat (at least at first) unconscious drive to reframe the collective possibilities within which we define excellence, achievement, and mastery of any particular discipline. They are here to break apart the consensus ideas of what constitutes success and then to redefine the responsibilities that come along with being a public figure. Their contribution to evolving the collective is to find new combinations and new symbols of success, mastery, and public service, which help reframe these concepts within their particular cultural milieu.

Delineating the pre-natal and post-natal inferior conjunctions by (whole sign) house:

Houses Ten to Two
The combination of the pre-natal inferior conjunction of Mercury and the sun located in the tenth house and the post-natal inferior conjunction in the second contains a deep (unconscious) drive to redefine the collective roles within which an individual may become an agent of the collective. Through their personal will, these people can move the collective in a new direction. Eventually, this will likely involve the discovery and exploitation of new resources, including intangible personal resources but also resources belonging to the collective, of which they become the steward. This stew-

ardship could involve being the holder of records, honors, and public acclaim, as well as literal physical capital.

Fire signs: Aries, Leo, and Sagittarius

The fire persona combines with the Mercury retrogrades in earth signs to form a potentially highly combustible mixture. These people can eventually find themselves in, or become, the heart of a firestorm. This combination is a natural transformation where the imaginative faculties can succeed in breaking apart previously restricting structures, and the energy released can become a tremendous inspiration. For instance, *Aries rising* native **Robert Watson-Watt** was a pioneer of radar implementation and designed systems that provided advance warning of approaching aircraft from a much greater distance. This vital intelligence proved invaluable in the Second World War, and Watt received high honors from multiple nations for his role in the Allied victory. For *Leo rising*, we can see the development of new forms of mastery in Dutch artist **M. C. Escher**, who studied the relationships between mathematics and nature and used tessellation to produce artwork that has become widely known in popular culture and influential in both art and mathematics. **Robert Duvall** (*Sagittarius rising*) developed mastery of the acting craft as a character actor early in his career, becoming known for his ability to completely immerse himself in character and transform into the role he's playing.

Air signs: Gemini, Libra, and Aquarius

The transformation of water into air is a natural ongoing process such that over time these individuals can gradually build a level of mastery that gains momentum and establishes an enduring contribution. The air identity has the natural ability to see things from a wider perspective, and the Mercury retrogrades in water signs give the additional ability to transport others to these higher places through shared emotional experience. For instance, *Gemini rising* native **Queen Victoria**'s reign coincided with a period of prosperity and expansion for the British Empire. She was a popular Queen who became a public symbol for family values, and after the posthumous release of her extensive diary and letters, the extent of her political influence became known. For *Libra rising*, we can see the development of new forms of mastery in **Alexandra David-Néel**, who became famous for entering the forbidden holy city of Lhasa, Tibet. She wrote over thirty books on Eastern religion, philosophy, and her travels, and these became particularly influential to the Beat writers of the '50s. **David Bowie** (*Aquarius rising*) developed a unique persona, and his innovations in glam rock images and new wave sounds helped bring new sophistication and intellectual depth to rock music. Bowie's four decades of artistic excellence were marked by perpetual reinvention and a continuously flowing but always unique, inspirational, and enduring image and sound.

Water signs: Cancer, Scorpio, and Pisces
The water persona in combination with the Mercury retrogrades in fire signs is a powerful but difficult transformation which requires complete immersion within a solid, stable structure or container to keep the polarized elements separate, and/or an intermediary step wherein the transformation is allowed to happen gradually in discrete steps or stages. This latter path can ultimately result in the arrival at a unique destination and public persona quite different and more controversial than the one originally sought. For instance, *Cancer rising* native **Mary Decker** demonstrated new forms of mastery by becoming the first woman to record a sub 4:20 time in the mile and helped rewrite track history by breaking the mile world record on three different occasions in the '80s. Her US national records for the mile, 2000 and 3000 meters still stand. For *Scorpio rising*, we can see these principles of developing new forms of mastery exemplified by **Cherrie Moraga**, an award-winning author who is one of the few writing about the intersectionality between multiple sources of marginalization, such as gender, sexuality, and race, in the experiences of women of color. **Alexander Graham Bell** (*Pisces rising*) experimented with electricity and sound as a result of his mother and wife's deafness. He narrowly managed to file the first patent for a practical telephone and faced many years of court battles challenging his company's patents, but ultimately prevailed due to his notes and letters establishing the long lineage of his experiments.

Earth signs: Taurus, Virgo, and Capricorn
The earth persona in combination with the Mercury retrogrades in air signs is a rather problematic transformation between polarized elements which may require an outside catalyst and/or intermediary step, wherein the transformation is allowed to happen gradually in steps or stages. The tremendous power ensuing from this blending of polar opposites can sometimes result in tragedy or controversy, commingled with success. For instance, *Taurus rising* native **Judith Resnik's** academic and professional excellence as an electrical engineer led to her being recruited by NASA. She later became the first Jewish-American astronaut and the first Jewish woman to go into space. She died tragically in the Space Shuttle *Challenger* disaster. For *Virgo rising*, we can see the principles of new forms of mastery exemplified by **Paul McCartney**, who initially gained fame as bass player and songwriter for the Beatles. McCartney's later career included much solo success as well as various other collaborations, making him one of the best-selling performers and composers of all time. **Dave Grohl** (*Capricorn rising*) initially rose to fame as the drummer for the genre-transforming band Nirvana. After the death of Nirvana founder Kurt Cobain, Grohl displayed multi-instrumental mastery as the front-man and principal composer of the Foo Fighters and has recorded and performed in numerous other bands.

Delineating the pre-natal and post-natal inferior conjunctions by (whole sign) house:

Houses Two to Six
The combination of the pre-natal inferior conjunction of Mercury and the sun located in the second house and the post-natal inferior conjunction in the sixth contains a deep (unconscious) drive to develop personal inspirations, talents, and resources in a way which redefines the possibilities of individual contribution to the collective. Often this involves development, extension, and mastery of new skill sets, of which the individual becomes personal exemplar. By further developing the example of their own personal influences

and applying the talents they discover via this emulation in new unexpected directions, they can become exemplars of a new form of virtuosity, which many others may strive to repeat and emulate, thus extending the waves of those who influenced them far and wide into the collective.

Fire signs: Aries, Leo, and Sagittarius
The dry fuel symbolized by the Mercury retrogrades in earth signs combusts easily for the individual with a fire sign rising. The natural dynamism of the fire element lends itself to heroic expression through which the individual learns to take on the cause of the masses or develops symbols for them. This collective responsibility can feel like a burden to some fire types and result in a tumultuous personal life filled with controversy or tragedy. For instance, *Aries rising* native **Pearl Buck** redefined mastery by becoming the first American woman to win the Nobel Prize for literature. She not only successfully managed the multiple careers of wife, mother, author, and editor but also contributed much as a political activist and advocate for the rights of women, children, and minority groups. Buck co-founded the first international, interracial adoption agency. For *Leo rising*, we can easily see the principles of new forms of mastery in the career of **Gustave Eiffel**, who gave us the iconic structure of the Eiffel Tower, which, as the world's highest structure at the time, was very controversial. Prominent artists and writers signed a petition and letter condemning the tower as an eyesore, and the controversy extended to Eiffel's involvement in a project to build a canal in Panama, leading him to retire from engineering. Eiffel went on to make significant contributions in aerodynamics and meteorology. **Marlon Brando** (*Sagittarius rising*) redefined mastery in acting through his embrace of method acting, which focuses on the actor experiencing and embodying the role rather than simply representing it. The heightened emotionality of his acting reflected itself in Brando's tumultuous and tragic personal life.

Air signs: Gemini, Libra, and Aquarius
The air persona in combination with the Mercury retrogrades in water signs can result in a natural emotional power and the ability

to cohere diverse elements into new forms with universal appeal, due to the quality of moistness and the ability to flow shared by both elements. This universality combines with the power of air to rise above the everyday, forming a transcendental movement which has the cohesive power to transport others beyond existing limitations. For instance, *Gemini rising* native **e e cummings** developed new forms of mastery by experimenting playfully and radically with poetic form while remaining true to timeless subject matter such as love, war, and sex. Cummings non-conformist style and political stances necessitated self-publishing and early financial struggles, but through the emergence of counter-culture in the '40s and '50s he achieved great popularity. For *Libra rising*, we can see these principles of new forms of mastery exemplified by **Richie Havens**, who achieved initial stardom by electrifying the crowd as the first performer at the historic Woodstock music festival. Havens made up for delays in other performers arriving with an extended, improvised set that culminated with a rendition of the old spiritual tune "Motherless Child," which morphed into a cry for "Freedom." His intense, raw, and rhythmic synthesis of various traditional styles led to lasting recognition. **Billie Holiday** (*Aquarius rising*) developed a distinctive jazz vocal style that made her famous and influential to later generations. As a songwriter, Holiday was unafraid of controversy and ahead of her time, most notably in her song "Strange Fruit" (1939), addressing the lynching of African Americans in the South.

Water signs: Cancer, Scorpio, and Pisces
The water persona in combination with the Mercury retrogrades in fire signs is a challenging transformation between polarized opposites which may require an intermediary step or strong container, but when successfully managed can harness tremendous power and intensity. Both water and fire can be more subjective and instinctive than rational or detached, so this combination may need an outside observer or objective sounding board to dial into the exact nature of its collective destiny. For instance, *Cancer rising* native **Jessye Norman** displayed new forms of mastery in operatic performance. Norman has taken her award-winning dramatic so-

prano voice in unusual directions, seeking out roles beyond the traditional limits of a particular range. For *Scorpio rising*, we see these principles of new forms of mastery evinced by **David Lynch** who became known for his enigmatic, disturbing, and surrealist cinematic style that explores the seedy underbelly of American culture and which became quite popular and influential to mainstream commercial tastes. **Herbert Mills** (*Pisces rising*) helped to redefine mastery by unusually expressing the human voice. By learning to mimic the sounds of a brass orchestra, The Mills Brothers became a highly successful jazz and pop vocal quartet who went far beyond a novelty act to enjoy decades of popularity.

Earth signs: Taurus, Virgo, and Capricorn

The combination of earth persona with Mercury retrogrades in air signs is another difficult combination which may require an intermediary step. The natural tendency toward detachment and objectivity of both elements can require an emotional catalyst, but once underway the inherent stability of both earth and air can hold the momentum and carry it forward. For instance, *Taurus rising* native **Jean Cocteau** lost his father at nine and left home at fifteen, becoming known in his twenties as the "frivolous prince" in the Bohemian social circles of Paris. Cocteau's experiments as a dramatist and filmmaker as well as his social connections had an enormous influence on the development of French avant-garde art movement. For *Virgo rising*, we can see the development of new forms of mastery in **Tiger Woods**, who radically changed the game of golf with his prodigious drive lengths and unprecedented endorsement deals. For several years Woods became one of the highest-paid athletes in the world, before enduring an equally spectacular fall from grace due to the publicizing of multiple marital infidelities. **Georgie Ann Geyser** (*Capricorn rising*) displayed new forms of mastery in journalism, becoming the first Western reporter to interview Saddam Hussein and authoring a biography of Fidel Castro. Her syndicated column appears in more than a hundred newspapers across both North and South America.

Delineating the pre-natal and post-natal inferior conjunctions by (whole sign) house:

Houses Six to Ten
The combination of the pre-natal inferior conjunction of Mercury and the sun located in the sixth house and the post-natal inferior conjunction in the tenth contains a deep (unconscious) drive to perfect an unusual talent, drawing from varied sources and inspirations to develop new skills toward achievement of status in a way which is outside the norm, breaking the mold, and redefining what mastery looks like. The person may be instinctively driven to break away from their initial training and redefine the basic building blocks and skill sets used within their profession, thus serving to redefine the profession itself.

Fire signs: Aries, Leo, and Sagittarius
The fire persona in combination with the Mercury retrogrades in earth signs is a natural transformation, yet also a highly combustible mixture that can result in significant controversy in addition to achievement. These individuals can combine diverse skills and talents in unusual ways that break traditional definitions of success and open new ground for those that follow. For instance, *Aries rising* native **Henry Miller** pushed the boundaries of mastery by innovating a style of semi-autobiographical novel with explicit language that also ultimately tested the limits of American laws regarding what is considered pornography by the mainstream. In 1964 the US Supreme Court decided his 1934 novel *Tropic of Cancer* was, in fact, literature, and this became an important moment regarding both freedom of speech and the sexual revolution. For *Leo rising*, we can see these principles of perfecting a new form of mastery powerfully evinced by **Howard Cosell**, who had a nose and appetite for controversy and ability for intellectual analysis more typical of "hard" news reporting which helped bring a sense of seriousness to sports reporting that was ahead of its time. Cosell's inimitable style helped usher in an era of smarter, more colorful broadcasters and 24/7 TV sports coverage. **Gladys Knight** (*Sagittarius rising*), displayed new forms of mastery as a "crossover" artist, able to score hits on mul-

tiple charts with different audiences and winning Grammy awards in three categories: R&B, Pop, and Gospel. Her eclectic multi-genre talent and success became an inspiration for later "fusion" sounds and bands.

Air signs: Gemini, Libra, and Aquarius

The air persona in combination with the Mercury retrogrades in Water signs is a natural, gradual transformation that can take universal, instinctual or emotional issues and lift them up, eventually reaching new heights for all the world to see. Bodily or emotional wisdom combines with intellect to produce a unique and transformative view of the human condition. For instance, *Gemini rising* native **RuPaul** perfected new forms of mastery by pushing the boundaries of gender awareness in the hit TV show *RuPaul's Drag Race*. For *Libra rising*, we can see the principles of new forms and levels of mastery embodied in one of the greatest basketball players of all time, **Kareem Abdul-Jabbar**, whose cleverly devised "skyhook" shot capitalized on his relatively slender frame, enabling him to evade blocks from defensive players. **Carl Gustav Jung** (*Aquarius rising*) took the still somewhat nascent field of psychology to new levels by making contributions in all "four forces" of the discipline. Jung's personal break with Freud can be seen to mirror a larger collective divide in the field of psychology itself. The pathos-centered and deterministic tendencies of both Skinner's behaviorism and Freud's psychodynamic theories gave way to a new focus on inherent health, growth, and spiritual aspiration in Jung's psychoanalytic theories, and this can be seen to parallel the emergent "third force" of humanistic psychology. Jung's focus on the collective unconscious and archetypes can be seen to prefigure the later rise of transpersonal psychology as the "fourth force."

Water signs: Cancer, Scorpio, and Pisces

The water persona in combination with the Mercury retrogrades in fire signs is a powerful but problematic transformation between polarized elements which can result in significant volatility. This combination requires a solid stable structure and/or intermediary steps or stages for the transformation to be complete. For instance,

Cancer rising native **Berthe Morisot** joined the controversial Impressionist movement of the late nineteenth century and later became known as one of its grand dames. Her mastery of color made her one of the highest selling female artists of modern times. For *Scorpio rising*, we can see these principles of new forms of mastery evinced by **Sylvia Porter**, who studied economics and finance after the Stock Market Crash of 1929 and eventually gained a readership of 40 million people. When readers discovered that the trusted "S. F. Porter" was a woman, her career developed even further into radio and magazines and as an advisor to President Lyndon Johnson. **Patricia McBride** (*Pisces rising*) made history early in life when she became the youngest principle dancer in New York City Ballet history. She went on to enjoy a storied thirty-year career with the company.

Earth signs: Taurus, Virgo, and Capricorn
The earth persona in combination with the Mercury retrogrades in air signs is a troublesome but powerful transformation which may require a catalyzing agent that can introduce significant volatility. It is best if the transformation happens gradually in steps or stages to offset the possibility of takeover by the volatile agent. For instance, *Taurus rising* native **Bob Hayes** first had a career as a successful Olympic sprinter and then went on to an NFL Hall of Fame career as an American football player. Hayes is the only man to win both a Super Bowl and an Olympic gold medal. For *Virgo rising* we can easily see the principles of new forms of mastery in the career of **Roy Orbison**, which peaked twice, once in the 1960s and then again in the 1980s, with a period of personal tragedy between them. Orbison created his unique style of rock and roll by combining complex classically styled composition with a soaring powerful, emotionally expressive, and operatic voice. **Buddy Holly** (*Capricorn rising*) most strikingly evinced new forms of mastery as a pioneer of rock and roll music who had an enormous influence on future generations. Holly displayed unusual artistic control and range for the time by not only writing but also recording and producing his material—a process later emulated by artists such as The Beatles. Holly's band The Crickets are also often regarded

as setting the standard for what became the traditional rock band lineup of two guitars, a bass, and drums.

THE RELATIONSHIP TRIGON: Houses Seven, Eleven, and Three

Because angular houses are more powerful by initiating important and essential activity, the angular house amongst any of these four sets defines the primary purpose of the entire trigon. The angular house of the 7/11/3 trigon is the seventh house, and so I call this the Relationship Trigon.

The angular house initiates important and essential activity; the succedent house serves to stabilize this purpose, by solidifying and holding the fruits of those activities; and the cadent house then serves to distribute these fruits, which in turn catalyzes further action. We can see then that the Relationship Trigon consists of three basic parts: our closest relationships with individuals most important to us (seventh house) which we take on in order to see and understand ourselves more completely, our intra-group relationships and the social capital we acquire within and through group membership (eleventh house), and the inter-group or community relationships we forge with those who share our daily environment and routines (third house).

Individuals who are born when the three Mercury retrogrades of their birth year activate these three houses have a deep, instinctive, and somewhat (at least at first) unconscious drive to reframe the collective possibilities within which we define relationships, of which there are three basic kinds: one to one, intra-group, and inter-group. They are here to break apart the consensus ideas of what constitutes healthy and successful relationship and the responsibilities and benefits that come with being a part of such. Their contribution to evolving the collective is to find new combinations and new symbols of successful relating which help to reframe these concepts within their culture.

Delineating the pre-natal and post-natal inferior conjunctions by (whole sign) house:

Houses Seven to Eleven
The combination of the pre-natal inferior conjunction of Mercury and the sun located in the seventh house and the post-natal inferior conjunction in the eleventh contains a deep drive to redefine the ways in which relationship expands and defines the collective contributions available to the individual. New ways of relating serve to redefine and expand the group memberships and social capital available to individuals. These people change society by changing and/or reframing the social connections and combinations through which people can make successful collective contributions.

Fire signs: Aries, Leo, and Sagittarius
The fire persona in combination with the Mercury retrogrades in air signs is a highly volatile mixture which will tend to separate out the subtle, higher, and more rarefied qualities from the coarse, lower, or more material aspects of any relationship. Thus, idealism can often become both a saving grace or downfall for these individuals and the defining force of their collective contribution. For instance, *Aries rising* native **Gary Hart**'s campaign for president in 1988 was scandalized by the media when allegations of womanizing and angry creditors surfaced. This event pointed to a growing adverse relationship between the media and politicians or celebrities and augured later scandals such as those involving Bill Clinton and Princess Diana. For *Leo rising*, we can see the principles of redefining collective relationships most clearly evinced in the figure of King **George III** of Great Britain, who oversaw a period of major military conflicts and shifting alliances. Ironically, King George was at once the personification of tyranny for the American revolutionaries and yet also a patron of Enlightenment scholarship. His library was one of the most important collections for its time and was open to scholars regardless of their political agreement, or lack thereof, with the King. **Martin Heidegger** (*Sagittarius rising*), widely considered one of the most influential philosophers of the twentieth century, focused on ontology (the question of being) and

its relationship to time and history. Heidegger saw a problem with true objectivity in Western philosophy and sought a way to recover a more primary notion of being beyond the pre-conceived notions (particularly of Aristotle) passed down through tradition. Paradoxically, Heidegger's questions about the more primordial natures of being and time were ultimately left unanswered, and yet his call for authenticity and use of deconstruction have themselves become pre-conceived notions within the development of existentialism and post-modernism.

Air signs: Gemini, Libra, and Aquarius
The air persona in combination with the Mercury retrogrades in fire signs is another highly volatile mixture which will tend to separate out the subtle, higher, and more rarefied qualities from the coarse, lower, or more material aspects of any relationship. Sudden or unexpected initiatory experiences into the lofty esoteric realms of fire should result in attempts at deeper understanding and recalibration of theoretical positions. For instance, *Gemini rising* native **Albert Hofmann** inadvertently transformed collective relationships when he synthesized and discovered the psychedelic effects of LSD. Hofmann's "problem child," as he later referred to it, would become a recreational drug that sparked a significant counter-culture but also served as a catalyst in the development of Transpersonal Psychology. For *Libra rising*, we can see the development of new symbols of relating in **Andy Griffith**, who rose to stardom based on his performance in the film *A Face in the Crowd*. Griffith gives a powerful performance as Larry Rhodes, a charismatic drifter who charms the public as a sudden TV sensation only to become drunk with his popularity and disparaging of his audience. Perhaps because of the archetypal nature of its story or the early challenge to the hypnotic power of TV it represents, the film has gained in popularity with critics over the years, and some have even come to see it as presaging the rise of media–turned–political figures like Donald Trump. **Sylvia Plath** (*Aquarius rising*) changed the way we think about relationships on several levels. Though initially hesitant about it, Plath became famous for a confessional form of poetry, wherein she brought the reader into relationship with her personal

experiences of subject matter not heretofore discussed openly. Further, Plath explored her relationship to love itself in poems such as "Love is a Parallax," and her stormy relationship with husband Ted Hughes has been a source of public fascination for decades, with her personal journals alternately scrubbed by Hughes and then more recently restored. Finally, Plath's fame necessitates a review of our collective relationship to mental illness, as she endured a long battle with depression, eventually succumbing to it and dying prematurely of suicide in 1963.

Water signs: Cancer, Scorpio, and Pisces
The water persona combines with the Mercury retrogrades in earth signs to create the possibility of boundary crossings of social barriers which ultimately serve to reshape the boundary itself in the long term. The relative density of these two elements can, however, make it difficult to separate the more subtle and refined effects from the baser aspects. For instance, *Cancer rising* native **Vanessa Williams** became the first black Miss America and then had to resign near the end of her term due to a nude photo scandal. She bounced back with a successful singing and acting career, which earned her a star on the Hollywood Walk of Fame. For *Scorpio rising*, we can see these principles in the career of **Ann Wilson**, who was a pioneer as one of the most outstanding female rock vocalists for the hard rock band Heart. Wilson's struggles with weight and substance abuse had a negative impact on her relationships and career with Heart, but she nonetheless accomplished a later solo career. **Warren G. Harding** (*Pisces rising*) rose to the highest office in the land as the popular 29th President of the United States and became the first president to visit Alaska, however numerous scandals resulting from his relationship choices have since enormously eroded his legacy. The Teapot Dome bribery scandal was probably the most sensational in US history, before Watergate, resulting in the first former cabinet member to receive a prison sentence and establishing that Congress has the right to compel testimony.

Earth signs: Taurus, Virgo, and Capricorn
The earth persona combines with the Mercury retrogrades in water signs to create a stable, dependable social identity that nonetheless has a tendency to make waves when teaming up with other independent thinkers. For instance, *Taurus rising* native **Jeanne Moreau** maintained friendships and relationships with prominent writers, directors, and musicians throughout her life. Her early film credits include work with many of the best-known New Wave and avant-garde directors, and she continued to make films into her eighties. For *Virgo rising*, we can see these principles of new social relationship evinced by **Joan Quigley**, who first met Nancy Reagan in the 1970s on *The Merv Griffin Show* and later went on to serve as an unofficial advisor to the Reagan White House in the 1980s after the assassination attempt on Reagan's life. Quigley moved in upper-class social circles and was an advanced astrologer, especially for her time, apparently benefitting from classical astrology training beginning at an early age with an experienced elder astrologer. She practiced mostly electional astrology for the Reagans, helping choose the timing of important events. **Jesse Jackson** (*Capricorn rising*) was part of the Civil Rights movement of the '60s, working for Dr. Martin Luther King, Jr. Later, Jackson expanded from race relations into international diplomacy, being instrumental in the release of American prisoners held in Syria and Cuba in 1983 and 1984. This success led to presidential bids in 1984 and 1988, where Jackson was seen as a fringe candidate and yet captured a surprisingly substantial portion of the vote, paving the way for other non-mainstream candidates.

Delineating the pre-natal and post-natal inferior conjunctions by (whole sign) house:

Houses Eleven to Three
The combination of the pre-natal inferior conjunction of Mercury and the sun located in the eleventh house and the post-natal inferior conjunction in the third contains a deep rooted, instinctive, and often somewhat (at least at first) unconscious drive to redefine

the ways in which group membership expands and defines the collective contributions available to the individual. New ways of relating within groups and exploiting social capital serve to redefine and expand the cultural milieus available for individuals to express themselves within. These people change society by reframing the social groups and cultural combinations through which people can make successful collective contributions.

Fire signs: Aries, Leo, and Sagittarius
The fire persona combines with the Mercury retrogrades in air signs to create a highly volatile mixture which will tend to force a separation of the coarse, lower, or more material aspects associated with any groups from the subtle, higher, and more rarefied qualities resulting from group membership. Thus, idealism can often become the defining force of their collective contribution. They seek to raise the group's status within the wider collective community. For instance, *Aries rising* native **Barbra Streisand** embodies the principles of pioneering new cultural combinations by having become the first woman to write, produce, direct, and star in the same film. Beyond being the only recording artist to have a number-one album in each of the last six decades, Streisand has demonstrated excellence across the entire range of performance genres. She is one of only two people to win top awards in television (Emmy), recording (Grammy), film (Oscar) and theater (Tony) as well as the distinguished Peabody Award. For *Leo rising,* we can see the principles of refinement of social group status tragically evinced by **Anne Frank**, who became one of the most well-known Jewish victims of the Holocaust. The posthumous publication of *The Diary of a Young Girl* had a lasting impact on the world, serving as inspiration for many, including Nelson Mandela in his struggle against apartheid. **Jimi Hendrix** (*Sagittarius rising*) illustrates the principles of new social group combinations by being an American who first gained recognition as a star in the London music scene and so became associated with the "British Invasion" of UK rock bands into the US during the '60s. His famous psychedelic reinterpretation of the US national anthem "The Star-Spangled Banner" played at dawn on the last day of the epochal Woodstock music festival,

challenges us to reconcile our national pride with our collective role in perpetuating anguish and destruction via military interventions (at the time in Viet Nam).

Air signs: Gemini, Libra, Aquarius
The air persona combined with the Mercury retrogrades in fire signs results in a volatile mixture that seeks to bring the more rarefied elements of a cultural identity into a more widely accessible form, such as through the mixing of classical with popular forms. For instance, *Gemini rising* native **Julia Child** heightened the awareness of French cuisine for the American public through popular television programs and cookbooks. For *Libra rising*, we can see the principles of new cultural combinations evinced by American chef **Emeril Lagasse**, who created a new style of cuisine by mixing Cajun and Creole styles with other cultural influences in a combination dubbed "New New Orleans." Lagasse popularized his innovation through his television shows, cookbooks, and restaurants as well as merchandising and endorsements. **Nina Simone** (*Aquarius rising*) recombined cultural identities by fusing a broad range of musical styles such as gospel, pop, and classical into a unique soulful blend, which also served as a vehicle for political activism and civil rights. These protests against civil injustice eventually led her to expatriate, and she lived in several countries before settling in the south of France and experiencing a resurgence in popularity during the '80s and '90s.

Water signs: Cancer, Scorpio, and Pisces
The water persona in combination with the Mercury retrogrades in earth signs is a natural transformation that can bring solidity and stability to volatile situations while at the same time slowly and subtly transforming them to include a more universal appeal. For instance, *Cancer rising* native **Frank Gifford** demonstrated these principles in his legendary twelve-season American football career with the New York Giants. Gifford was an all-around athlete and one of the last to play "both ways" on offense and defense, chalking up impressive statistics in almost every phase of the game. Gifford then went on to provide a low-key counter-balance to the volatile

Howard Cosell and Don Meredith as the play-by-play announcer for *Monday Night Football*. He called almost 600 games over 27 years, becoming a sports icon and helping transform *Monday Night Football* into an American institution. For *Scorpio rising*, we can see these principles of long-term social transformation evinced by **Tom Bradley**, who became only the second African-American mayor of a major city, and his twenty-year tenure as Mayor of Los Angeles is the longest for LA. Bradley was able to combine a pro-business but politically liberal agenda that was ahead of its time, signing the city's first homosexual rights bill in 1979 and anti-AIDS-discrimination bill in 1985. **Bette Nesmith Graham** (*Pisces rising*) was an artist and secretary who invented the transformative agent commonly known as "liquid paper" to more easily correct typing errors. As a businesswoman, she sought to create a more nurturing business environment by integrating spirituality, egalitarianism, and pragmatism into the workplace.

Earth signs: Taurus, Virgo, and Capricorn

The earth persona in combination with the Mercury retrogrades in water signs is a natural transformation that brings formidable strength and staying power to the natural relentless dynamism of water. These people can become personal containers for social change. For instance, *Taurus rising* native **Charley Pride** embodies the principles of change within and between social and cultural groups by having become one of the few African-Americans inducted into the Grand Ole Opry. Pride's popularity as a country musician later helped him to break an effective concert ban in Northern Ireland, and his influence there still endures. For *Virgo rising*, we can see these principles of lasting social change lovingly evinced by **Eunice Kennedy Shriver**, founder of the Special Olympics. Shriver served as a lifelong advocate for children's health and disability issues and is the only woman to make an ante mortem appearance on a US coin. **Constance Markievicz** (*Capricorn rising*) was born to a wealthy family in London and became a countess by marriage, but following her father's concern for the poor, traded the privileged life of a socialite for the down-and-dirty world of political activism and guerilla warfare. Having been involved in the violent

Easter Rising rebellion, she nonetheless managed to become the first woman elected to the British House of Commons while incarcerated. She also became one of the first women in the world to hold a cabinet position, as Minister for Labour of the Irish Republic, 1919–1922.

Delineating the pre-natal and post-natal inferior conjunctions by (whole sign) house:

Houses Three to Seven
The combination of the pre-natal inferior conjunction of Mercury and the sun located in the third house and the post-natal inferior conjunction in the seventh contains a deep rooted, instinctive, and often somewhat (at least at first) unconscious drive to redefine the ways in which the inter-group relationships amongst the various cultural milieus navigated on a daily basis define the collective contributions available to an individual. There is a strong urge to break free from any dogmas and taboos restricting access to diverse input and to forge a closer personal relationship to a newly emergent cultural form, which usually represents a novel mixture of elements. These people change society by reframing the cultural combinations within which an individual may forge social identity.

Fire signs: Aries, Leo, and Sagittarius
The fire persona in combination with the Mercury retrogrades in air signs is a natural transformation but nevertheless a combustible mixture. The highest and most subtle parts of the nature need conceptualization within a standard cultural form to be identifiable, and yet this can sometimes bring a backlash from the coarser side, involving transgressions against or limitations by societal strictures or taboos. For instance, *Aries rising* native **Yo-Yo Ma** has not only had a prolific international career as a classical cellist but has also recorded a wide variety of international folk music and collaborated with a range of artists from varied genres and mediums. For *Leo rising*, we can see these principles of reframing social identity most clearly evinced by **Charles, Prince of Wales**, who has always

been an independent thinker, promoting environmental awareness, organic farming, and alternative medicine as well as showing non-mainstream philosophical and religious interests. Charles also set new royal relationship standards more closely aligned with the times by being present at his children's births and also by divorcing and re-marrying in a civil, rather than religious, wedding. **Marianne Alireza** (*Sagittarius rising*) became the first western woman to live in Saudi Arabia, where she stayed for twelve years in a traditional harem. She chronicled the famous adventure in her autobiography *At the Drop of a Veil*.

Air signs: Gemini, Libra, and Aquarius

The air persona in combination with the Mercury retrogrades in fire signs is a natural transformation which seeks to draw out the more subtle rarefied qualities, while also rendering them in a more widely recognizable format. This combination is a hard-to-contain volatile mixture, but if successfully integrated can result in significant cultural advances. For instance, *Gemini rising* native **Patsy Cline** was raised by a single mother and yet managed to rise from her poor, working-class roots to become one of the first country singers to successfully crossover to pop music. A hard-working performer on the local circuits, she was eventually "discovered" via the television show *Arthur Godfrey's Talent Scouts*, an early version of modern shows like *American Idol*. Cline became one of the most influential, successful, and acclaimed vocalists of the twentieth century, helping pave the way for generations of women as headline performers. For *Libra rising*, we can see these principles of cultural recombinations most clearly evinced by **Yoko Ono**, who is well known for her avant-garde work in a wide range of mediums including performance art, music, and filmmaking—in addition to her collaboration as peace activists with her late husband, John Lennon. **Marguerite de Navarre** (*Aquarius rising*) was an outstanding figure of the French Renaissance, both as an author and patron of humanists and reformers. Together with her brother, King Francis I, she presided over the celebrated intellectual and cultural court and salons of their day.

Water signs: Cancer, Scorpio, and Pisces
The water persona in combination with the Mercury retrogrades in earth signs is a natural transformation that seeks to condense new forms of emotional awareness into physical forms. The relationship to the physical world becomes a metaphor for a higher conceptual awareness based on an emotional response to an ideal. For instance, *Cancer rising* native **Paul Delvaux** developed a unique surrealistic style of painting that juxtaposed classical perfectionism with bizarre combinations of erotic and morbid subject matter designed to induce emotional shock in the viewer. Delvaux's compositions of female nudes, skeletons, and classic architecture—all rendered with perfect realism—evoke a hallucinatory mood by way of the incongruity of their relationships to normal experience. For *Scorpio rising*, we can see these principles of new inter-group relationships most clearly in the work of French Post-Impressionist painter **Paul Cezanne**, who radically altered the course of twentieth-century art. Cezanne formed close relationships with other important artists early in his career, most notably with the writer Émile Zola and later with the Impressionist painter Camille Pissarro. After a period of social and artistic isolation, Cezanne slowly developed his unique explorations of geometric simplification and optical phenomena, which inspired the advent of other styles, such as Cubism, and paved the way for the abstract art of the twentieth century. **Peter Yarrow** (*Pisces rising*) demonstrated a unique blend of cultural relationships as part of the folk music explosion of the early '60s. He met the musical manager and impresario Albert Grossman through his performance at the Newport Folk Festival, and this resulted in the formation of the very popular trio Peter, Paul, and Mary. Their performance of fellow Grossman client, Bob Dylan's song "Blowin' in the Wind" at the Reverend Martin Luther King's historic March on Washington established the song as a civil rights anthem.

Earth signs: Taurus, Virgo, and Capricorn
The earth persona combines with the Mercury retrogrades in water signs in a natural transformation that raises the sensate function or bodily awareness into the emotional sphere and seeks to con-

vey this new blend in terms of budding cultural relationships. For instance, Brazilian author and *Taurus rising* native **Paulo Coelho** demonstrated the principles of transformation via cross-cultural relationships by taking up a pilgrimage on the Camino de Santiago in Spain, where he experienced a spiritual awakening. Afterward, he transformed from a lyricist into a novelist, penning *The Alchemist* the following year, which has since become the most translated book by a living author. For *Virgo rising*, we can see these principles of emergent cultural forms most clearly evoked by **Kurt Cobain**, who helped forge a new genre of rock music called "grunge," which fuses elements of punk rock and heavy metal. Cobain was co-founder and principal songwriter of the band Nirvana, whose signature sound developed out of the use of strong contrasts between fast and slow, loud and quiet. This use of contrasts is most apparent in the epoch-defining song, "Smells like Teen Spirit," whose structure mysteriously invokes the same pattern as the star of Mercury (see Chapter One), with six verses, alternating between lingering and racing, soft and heavy, fixed and volatile. **Clarence Thomas** (*Capricorn rising*) inadvertently brought new cultural relationships into focus, when his confirmation hearings for Justice of the Supreme Court of the United States brought the issue of sexual harassment in the workplace to the fore. This spotlight, in turn, led to both the election of a record number of women to office in the following years and companies establishing or modifying sexual harassment policies.

The Trigon of Dynasty: Houses Four, Eight, and Twelve

Given that angular houses are more powerful by initiating important and essential activity, the angular house amongst any of these four sets defines the primary purpose of the entire trigon. The angular house of the 4/8/12 trigon is the fourth house, and so I have termed this the Trigon of Dynasty because traditionally our family ties are both where we come from and what we ultimately leave behind (the end of the matter).

The angular house initiates important and essential activity; the succedent house serves to stabilize this purpose, by solidifying and holding the fruits of those activities; and the cadent house then serves to distribute these fruits, which in turn catalyzes further action. The Trigon of Dynasty, therefore, consists of three basic parts: the group with whom we share the most biological similarity (fourth house) and the genetic inheritance which we take on (and leave behind) in order to incarnate; our marital family groups and the material capital we acquire through marriage ties and/or extended family (eighth house); and the spiritual family groups we forge with those who share our broader spiritual ideas and ideals (twelfth house).

Individuals who are born with the three Mercury retrogrades surrounding their birth activating these three houses are incarnating with a deep-rooted, instinctive and somewhat (at least at first) unconscious drive to reframe the collective possibilities within which we define family, of which there are three basic kinds: birth, marital/contractual, and spiritual. They are here to break apart the consensus ideas of what constitutes healthy and successful family and the responsibilities and benefits that come with being a part of such. Their contribution to evolving the collective is to find new combinations and new symbols of successful familial groups and dynastic contributions, which help to reframe these concepts within their culture.

Delineating the pre-natal and post-natal inferior conjunctions by (whole sign) house:

Houses Four to Eight
When the pre-natal inferior conjunction of Mercury and the sun is located in the fourth house (and thus the post-natal inferior conjunction usually in the eighth) there is a deep (unconscious) drive to redefine how family membership expands and defines the collective contributions available to the individual. These people may experience ambivalence in their emotional ties to their biological family of origin or national roots, perhaps even when there is no specific behavior or reason that can explain or justify these feelings. They will instinctively seek changes via the shared resources of their marital group memberships or other close partnerships.

Finding and exploiting new emotional capital within the marital, partnership, or extended family group memberships will eventually result in a natural process of adjustment to their dynastic prospects.

Fire signs: Aries, Leo, and Sagittarius
The fire persona combines with the Mercury retrogrades in water signs to create a highly volatile mixture. An intermediate stage of transformation and/or strong container is often necessary here, but even then the mixing of polarized opposites can sometimes result in uncontrolled or even violent reactions. For instance, *Aries rising* native **John Lennon** reveals the principles of redefining family membership in several ways. Lennon rose from working-class roots in a broken family to worldwide fame as a co-founder of the Beatles, who revolutionized virtually everything about popular music and opened the door for a tidal wave of rock acts known as the British Invasion. Lennon married Yoko Ono in 1969, The Beatles subsequently disbanded, and Lennon's relationships with his band-mates, ex-wife, and son Julian became strained. However, this new creative partnership with Ono resulted not just in further musical innovation (earning Lennon another membership in the Rock and Roll Hall of Fame) but also in Lennon embracing his role as a father and husband as well as the flowering of his spiritual and philosophical legacy as a peace activist. For *Leo rising,* we can see the development of these dynastic principles through the career of **Flip Wilson**, who was the first black comedian to host a successful variety series. Through the tremendous popularity of his show, Wilson was able to support the careers of many up-and-coming artists, including Richard Pryor and George Carlin, and this focus on comedy helped break ground for future iconic shows such as *Saturday Night Live*. **Dane Rudhyar** (*Sagittarius rising*) left his native France and culture of origin to travel to the "New World" of America and become what he called a "seed-man," an individual who consciously acts as catalyst and information carrier in the life of a new cycle or phase of history for the collective. Accordingly, Rudhyar went on to pioneer humanistic and transpersonal astrology, leaving behind many now-classic books on astrology, spirituality, and culture.

Air signs: Gemini, Libra, and Aquarius

The air persona in combination with the Mercury retrogrades in earth signs is another volatile mixture of opposites; however, if care is taken to ground the new ideas in firmly rooted traditions then the volatility can be successfully contained and an enduring legacy secured. For instance, *Gemini rising* native **Bobby Short** anchored a spot as featured performer at New York's famous Café Carlyle for over 35 years. He dedicated his sixty-year musical career to promoting what he termed the "Great American Song," and was eventually recognized as a Living Legend by the US Library of Congress in 2000. For *Libra rising*, we can see the tenet of redefining family from a cultural or collective perspective quite clearly in the work of **Joseph Campbell**, who was a strong believer in the psychic unity of humankind, and so interpreted myth symbols like collective dreams. Campbell's re-contextualization of the many particular manifestations of human culture as emanating from one common source, for example in *Masks of God*, became very popular and deeply influenced further generations. **Emmeline Pankhurst** (*Aquarius rising*) fought for collective unity as a women's rights activist and founder of the Women's Social and Political Union (1903). Though she was unafraid to employ militant methods and even served jail time, Pankhurst nonetheless remains a celebrated figure in the women's suffrage movement.

Water signs: Cancer, Scorpio, and Pisces

The water persona in combination with the Mercury retrogrades in air signs is a natural transformation wherein the universal life-giving nature of water receives the heat of passion and can then be distributed widely throughout the social fabric. At the same time, the intellectual is brought down to an emotional vibration, making it more widely accessible and humanitarian. For instance, *Cancer rising* native **William Blake** became a seminal figure in Romanticism, which is characterized by emotive and personal experience, in contrast to the impersonal rationality of the Enlightenment. Blake notably used illuminated plates to enhance his poetry, and this admixture of image with word evokes a powerful experience, such that Blake's work continues to grow in influence over time, inspiring new generations and firmly establishing his legacy in popular culture. For *Scorpio rising*, we can see these precepts of unity through connection to collective roots in the career of **Aretha Franklin**, whose soulful performances embody the spiritual ability to transform oppression and anguish into vital beauty that inspires. As a top-selling artist and first female inductee into the Rock and Roll Hall of Fame, Franklin is a legend in her own time. Feminist pioneer **George Sand** (*Pisces rising*) often hosted notable artists and writers at her historic house and estate in the village of Nohant, France. The location also served as the setting for much of her writing, which displayed intellectual rigor and yet retained sensitivity for the rural poor, working classes, and women's rights.

Earth signs: Taurus, Virgo, and Capricorn
The earth persona combines with the Mercury retrogrades in fire signs to create a natural transformation wherein the solid stable nature of earth provides the combustible fuel whereby the individual can influence the collective by allowing themselves to burn their individual passions and spiritual presence into the collective. For instance, *Taurus rising* native **Vivien Leigh** portrayed a dying dynasty through her role as the heroine in *Gone With The Wind*.[10] As Scarlett O'Hara, she has to learn to take care of herself and extended family, even as the Civil War destroys the privileged lifestyle in which she was raised, casting Southern society into ruin. For *Virgo rising*, we can see the principles of redefining home, family, and heritage evinced by **Dolly Parton**, who grew from her country roots, crossing over to pop music and film, all while becoming a global superstar. Parton has heavily reinvested in her native East Tennessee community, bringing literacy and jobs to the economically depressed region, and her children's literacy campaign has grown to reach millions worldwide. **Elizabeth II** of the United Kingdom (*Capricorn rising*) was not expected to be Queen until the abdication of Edward VIII. She has gone on to become the longest-reigning and most traveled British monarch, during a time of tremendous changes, including decolonization, absolute primogeniture (women have the same rights to the throne as men), and improved relations with Ireland.

Delineating the pre-natal and post-natal inferior conjunctions by (whole sign) house:

Houses Eight to Twelve
The combination of the pre-natal inferior conjunction of Mercury and the sun located in the eighth house and the post-natal inferior conjunction in the twelfth contains an instinctive drive to redefine how relationship membership, sexual identity, and shared resources serve to define the collective contributions available to the individual. These people are here to make fundamental updates and changes to the image and nature of how an individual can embody an archetype.

Fire signs: Aries, Leo, and Sagittarius
The fire persona in combination with Mercury retrograde in water signs is a volatile combination of opposites, which requires a strong container and/or multiple steps or stages of transformation. It may take some time, repetition, or hindsight to distill the subtle spirit of the archetypal changes being channeled. For instance, *Aries rising* native **Joan Baez** challenged the political and social climate of the '60s by exalting the medium of "protest songs" from simple social commentary into an art form. While others like Bob Dylan moved on, Baez displayed a lifelong commitment to political and social activism, recently employing her skills for social resistance in a protest song about Donald Trump. Her induction speech for the Rock and Roll Hall of Fame concluded with a timeless plea for compassionate social change. For *Leo rising*, we can see the tenet of redefining an archetype embodied by **Maya Angelou**, whose award-winning literary career transformed the genre of autobiography. Angelou's contributions also include her work in Civil Rights, and as a respected spokesperson for black people and black culture, but perhaps her greatest legacy is that of a fearless survivor whose triumphs over racism, sexism, and sexual abuse continue to inspire generations. **Raquel Welch** (*Sagittarius rising*) helped transform the stereotype of female sex symbols as helpless blue-eyed blondes. By starring in Westerns as a vengeful gunslinger and other movies with physically demand-

ing roles and challenging themes, Welch helped to redefine cinematic portrayals of women.

Air signs: Gemini, Libra, and Aquarius
The air persona in combination with the Mercury retrogrades in earth signs is a complex and arduous combination of opposites that requires a spark of volatility to begin the transformation, but this can become difficult to manage and contain if the individual is not careful to regulate the levels of instability or is not careful about letting off steam. For instance, *Gemini rising* native **Sam Cooke** displayed unusual cleverness and range by founding both a record label and a publishing company and managing both the musical and business sides of his musical career. Cooke died under mysterious circumstances, long before his time, but his influence as both singer and composer lives on, as he contributed to the rise of a whole generation of soul music pop stars. For *Libra rising*, we can see these principles of archetypal renewal powerfully evinced by **Courtney Love**, who founded Hole, one of the best-selling rock bands ever fronted by a woman. Love has seen more than her fair share of controversy, yet some of her self-destructive antics serve as fuel for art—e.g., she was molested by an audience after stage diving and wrote the song "Asking For It," comparing it to rape. The defiant Love continued to stage dive and managed to redefine rock stardom through her powerful lyrics, which opened space for a female point of view. **Odetta Holmes** (*Aquarius rising*) originally trained in opera and theater, and later became an influential part of the American folk music revival of the '50s and '60s, inspiring many other famous artists, including Harry Belafonte, Joan Baez, and Janis Joplin.

Water signs: Cancer, Scorpio, and Pisces
The water persona in combination with the Mercury retrogrades in air signs is a natural transformation wherein the universal nature of water is warmed to transcend boundaries and distribute it more widely. This inspiration can take many forms, but the upward striving should eventually balance with the return of the life force back to the earth and the people. For instance, after becoming the first

southern black female and lesbian elected to the US Congress, *Cancer rising* native **Barbara Jordan** further transformed the role of the politician by delivering notable political speeches remembered as some of the best of the twentieth century. For *Scorpio rising,* we can see the transformation through relationship status and archetypal imagery most clearly evinced by **Jacqueline Kennedy Onassis**, who became a fashion icon as First Lady, and the first to have a press secretary. Her major project as a First Lady was to restore the White House's historical character, and she further instituted the continuing care of this history. **Phil Donahue** (*Pisces rising*), transformed the archetype of talk show host by being the first with a format including audience participation, and his path-breaking eponymous show enjoyed the longest continuous run of any syndicated talk show in US history.

Earth signs: Taurus, Virgo, and Capricorn

The earth persona in combination with the Mercury retrogrades in fire signs is a natural transformation where the solidity of earth provides fixed fuel for the volatile questing spirit of fire. This combination is difficult to contain or direct, but with hard work, repetition, and the proper sources of inspiration can burn a lasting clearing in the wilderness of history. For instance, *Taurus rising* native **Rodney King** became a national symbol for injustice in America due to the brutal beating he received from Los Angeles Police Department officers following a high-speed car chase. A citizen videotaped the arrest from a nearby balcony, sent it to a television station, and it was soon broadcast across the nation. The 1992 Los Angeles riots sparked when a mostly white jury initially acquitted the officers. For *Virgo rising,* we can see the embodiment of archetypal change boldly evinced by **Ernest Hemingway**, whose simple, economical, and understated writing style became iconic and greatly influenced the development of twentieth-century American literature. Hemingway was a minimalist and pioneered the use of omission, with most of the crucial story developments only being implied. **Nicholas Culpeper** (*Capricorn rising*) transformed the practice of medicine by publishing, both cheaply and in vernacular English, self-help medical guides of inexpensive herbal remedies for use by the poor. Culpeper's extensive cataloging of hundreds of medicinal herbs and systematization of the use of herbal medicines represented important development toward the evolution of modern pharmaceuticals, and his works have been in print continuously since the seventeenth century.

Delineating the pre-natal and post-natal inferior conjunctions by (whole sign) house:

Houses Twelve to Four

The combination of the pre-natal inferior conjunction of Mercury and the sun located in the twelfth house and the post-natal inferior conjunction in the fourth contains a deep-rooted, instinctive, and often somewhat (at least at first) unconscious drive to redefine the ways in which spiritual family membership and individual identification as an archetypal symbol expands and defines the collective contributions available to the individual. These people may experience ambivalence in their emotional ties to their spiritual family and will instinctively seek to make changes to the resources available to those who take on the burden of becoming a collective symbol. These people are here to break apart previous conceptions of people as icons and to put the pieces back together into a reformulated sense of the collective family of humanity.

Fire signs: Aries, Leo, and Sagittarius

The fire persona in combination with the Mercury retrogrades in water signs is a volatile mixture of opposites that requires a strong container and/or an intermediate transition for the transformation to be completed. Due to the high volatility involved, even a successful transformation and established legacy will likely not be without its controversial aspects, but this is also what makes it revolutionary. For instance, *Aries rising* native **Pete Rose** was a phenomenal baseball player, with unparalleled achievements in multiple phases of the game; however, his (at first) unapologetic admission of having gambled on games he was involved in led to his permanent ineligibility from the sport, including its hall of fame. For *Leo rising*, we can clearly see the transformation of iconic status evinced by **Pablo Picasso**, who despite much personal and professional controversy is nonetheless remembered as one of the greatest and most influential artists of the twentieth century who courageously, radically, and continuously rearranged ways to see the world. **Oprah Winfrey** (*Sagittarius rising*) rose from humble roots to become an American media icon and the first female African-American billionaire. Oprah's confessional style and brutal honesty about her personal struggles have made her a symbol of both survival and also the rise of a more inclusive American dream.

Air signs: Gemini, Libra, and Aquarius

The air persona in combination with the Mercury retrogrades in earth signs is a difficult transformation between opposites, which requires a catalyzing agent and a strong container or intermediate stages in order to complete the transformation. There is a strong urge to separate the higher, more subtle expressions from the gross, and yet a need to retain enough grounding to ensure reaching the other side. For instance, *Gemini rising* native **Neil Armstrong** achieved iconic status as the first human being to walk on the surface of the moon. Armstrong was also an important supporter of a human mission to Mars, and a vocal critic of abandoning human exploration of space beyond Earth orbit. For *Libra rising*, we can see these transcendent tendencies embodied by **Marilyn Bell Di Lascio**, who at just sixteen years old became the first person to swim across Lake Ontario, and was also the youngest person ever to swim the English Channel. **Piet Mondrian** (*Aquarius rising*) was influenced by Theosophy and Anthroposophy to seek to distill in his art the fundamental spiritual truths of nature through the simplest geometric forms. His De Stijl school and neo-plastic style consisting primarily of straight black lines, rectangles, and primary colors continues to be influential with artists, designers, and architects.

Water signs: Cancer, Scorpio, and Pisces

The water persona in combination with the Mercury retrogrades in air signs is a natural transformation wherein the universal nature of water is warmed and made less dense, revealing and distributing its healing properties more widely and thus encompassing and bringing together a larger spiritual family by way of a shared archetypal inspiration. For instance, *Cancer rising* native **Tenzin Gyatso**, Fourteenth Dalai Lama, and Nobel Laureate, was forced to flee his native Tibet after invasion by the Chinese. However, he has subsequently traveled the world, sharing his "religion of kindness" and teachings which bring people together on a broad range of human issues. For *Scorpio rising*, we can see these principles of widespread inspiration in **Edith Piaf**, the unofficial national chanteuse of France. Despite a tumultuous personal life and premature death, Piaf's legacy as one of the greatest performers of the twentieth century has been assured by several biographies and films. Award-winning film director, **Frank Capra** (*Pisces rising*) often portrayed an idealized portrait of American life, "rags to riches" stories told with compassion and concern for others held as a noble virtue. Capra's string of acclaimed films in the 1930s and 1940s made him influential with later generations.

Earth signs: Taurus, Virgo, and Capricorn
The earth persona in combination with the Mercury retrogrades in fire signs are a natural transformation, wherein the personal ego solidity serves as fuel for collective inspiration. Through simple repetition or by defending, delineating, or understanding limits these individuals can help redefine the roles of the archetypal person as well as the spiritual family. For instance, *Taurus rising* native **Casey Kasem** became a national fixture via his weekly radio music countdown program *American Top 40*, hosting it for almost two decades straight in the '70s and '80s, and again from the late 90's into the new millennium. For *Virgo rising*, we can see these principles in their tragic form as **Francis Farmer** became a symbol of persecution and mistreatment of psychiatric patients. She was arrested after failing to pay the balance on a traffic ticket, was then involuntarily committed and subjected to insulin shock treatments that produced comas. Many today see this treatment, and others like it, as violations of basic human rights. **Ralph Abernathy** (*Capricorn rising*) grew up on a farm, and his easy connection

to rural, poor, and working-class people was a natural compliment to his collaborator Dr. Martin Luther King Jr.'s more intellectual nature. Together, Abernathy and King formed a magnetic pair and achieved much needed social change in the American Civil Rights Movement.

Delineating the pre-natal and post-natal inferior conjunctions by different trigons

This section discusses the exceptional cases when the pre-natal inferior conjunction is in one element and the post-natal inferior conjunction is in a different element. In addition to the following material, there is another example of the transformation from one element to another at the end of chapter four. Chapter five deals with the topic of transformation between elements at length, so the reader is encouraged to consult those sections as well.

The shift between elements represents a significant turning point in the cycle of the Mercury elemental year, and it thus offers the possibility of remarkable shifts in both one's personal life and karma as well as the collective dharma and evolution of possibilities for humanity. On an individual level, people born during these shifts will have two different trigons activated in their chart, one by the pre-natal conjunction and another by the post-natal conjunction. There are four basic permutations of the trigon combinations to be activated. In each case, the most powerful combination comes from two angular houses being activated.

Identity becoming Dynasty: (pre-natal inferior conjunction in houses 1/5/9 and post-natal inferior conjunction in houses 4/8/12)

These individuals will be instinctively drawn toward the personal expression of a newly emergent image or activity in such a way as to become an exemplar or heroic figure of this burgeoning collective possibility. Their accomplishments in bridging two worlds may be especially timely, such that they become widely identified with it. This transition may eventually prompt them to become more concerned with collective issues than individual pursuits and accomplishments, often leaving behind a legacy far beyond the original breakthrough which brought them into collective awareness.

American aviation pioneer, **Charles Lindbergh** (Scorpio rising) was the first person to complete a solo transatlantic flight from New York to Paris. Lindbergh's pre-natal conjunction was at 12 Scorpio, in the water element, and his post-natal conjunction was at 29 Aquarius, in the air element. The symbolic transition from water to air is embodied in Lindbergh's signature accomplishment of transatlantic flight.

Mastery becoming Identity: (pre-natal inferior conjunction in houses 10/2/6 and post-natal inferior conjunction in houses 1/5/9)
These individuals will be instinctively drawn toward the mastery of an emergent professional activity or public role. If successful, this mastery will result in their being identified as an exemplar or heroic figure of this burgeoning collective possibility. Their accomplishments in bridging two worlds may be especially timely, such that they become widely identified with it. This transition may eventually prompt them to become more concerned with personal issues or truths, especially those that fly in the face of collective belief or tradition, often leaving behind a legacy of individual accomplishment that may help usher in a new era for their profession or public position.

Tycho Brahe (Aquarius rising) was a Danish nobleman, astronomer, astrologer and alchemist. Brahe's pre-natal conjunction was at 6 Scorpio, the tenth house from Aquarius, and his post-natal conjunction was at 24 Aquarius, in the first house. Always dedicated to empiricism, Brahe worked without telescopes to produce accurate and comprehensive astronomical observations. He refuted the Aristotelian conception of the cosmos as immutable heavenly spheres, and his work paved the way for the scientific revolution. Brahe's success as a scientist also greatly depended on political skill to obtain patronage and funding.

Relationship becoming Mastery: (pre-natal inferior conjunction in houses 7/11/3 and post-natal inferior conjunction in houses 10/2/6)
These individuals will be instinctively drawn toward the personal expression of a newly emergent relationship status in such a way as to bring the social contract implied by that relationship to further completion or fulfillment. Their accomplishments in bridging two worlds may be especially timely, such that they become widely visible along with their cause. This transition may eventually prompt them to become more concerned with collective attainment of an ideal than individual pursuits and accomplishments, causing them to leave behind an enduring sense of the need for collective improvements beyond the original breakthrough in relationships which brought them into collective awareness.

Martin Luther King, Jr. (Taurus rising) was a Baptist minister, leader in the American Civil Rights Movement, political activist, and humanitarian. King's pre-natal conjunction was at 1 Scorpio, the seventh house from Taurus, and his post-natal conjunction was at 18 Aquarius, in the tenth house. King's early message was that the long history of injustices perpetrated against black Americans was glaringly at odds with the original founding ideals of the country, and that this represented a breach of the social contract which had to be remedied. After his initial success in achieving major political reforms for the black community, King turned the power of his non-violent activism toward injustices such as war, poverty, and worker's rights.

Dynasty becoming Relationship: (pre-natal inferior conjunction in houses 4/8/12 and post-natal inferior conjunction in houses 7/11/3)
These individuals will be instinctively drawn toward rejecting the traditional status of their birth family or other social inheritance and replacing this with the personal expression of a newly emergent relationship status in such a way as to bring the wider relationship of individual and collective into question on many levels. Their

accomplishments in bridging two worlds may be especially timely, such that they become widely visible along with their personal causes. This transition may eventually prompt them to become more concerned with collective attainment of freedom within relationship than individual obligation toward tradition, so that they leave an enduring sense of the need for collective improvements beyond the original break from tradition which brought them into collective awareness.

Patron, activist, and humanitarian, **Diana, Princess of Wales** (Sagittarius rising) was a British noble by birth and member of the British royal family by marriage to Prince Charles. Diana's pre-natal conjunction was at 6 Cancer, the eighth house from Sagittarius, and her post-natal conjunction was at 30 Libra, in the eleventh house. Diana used her celebrity status to advocate for victims of leprosy and HIV/AIDS, to campaign for animal protection, and to fight against the use of landmines. Her high-profile divorce from Prince Charles can be seen to symbolize the universal problems associated with the social versus personal aspects of marriage at a time when the institution of marriage itself was becoming a central question worldwide. Her tragic death also brought to light the unhealthy relationship between celebrities and paparazzi.

SECTION II
The Mercury Elemental Year as a Transformative Process

Chapter III
Animism, Magic, and Myth

> *We must invent, or re-invent, a sustainable human culture by a descent into our pre-rational, our instinctive resources [...] What is needed is not tran-scendence but in-scendance, not the brain but the gene.*
>
> —THOMAS BERRY
> THE DREAM OF THE EARTH[11]

ANTHROPOLOGISTS SUGGEST IT was a "creative explosion" of primal art, such as cave paintings and figurines, which formally marks the transition to what we consider to be modern humans during the upper Paleolithic period. So, it is first our image- and later our symbol-making capacities which stand out as distinctively modern human adaptations and characteristics. Similarly, we have recently seen a tsunami of images in the digital age. The amount of images we now make is so unprecedented that it can scarcely be appreciated. As of publication of this book, more than 350 million photos are uploaded per day to the most popular photo sharing website and that number continues to grow. And that's just one website. Clearly image making is a timeless and fundamental human urge, and yet it is possible that the recent proliferation has in turn somewhat obscured the enormous magical and transformational power contained in an image.

In the 2010 documentary "Cave of Forgotten Dreams" a researcher is interviewed confessing that the lions painted on the walls of Chauvet cave had invaded their dreams with such a powerful and profound presence that they had to stop going inside the caves in order to process the profound emotions stirred by the experience.[12] This encounter truly exemplifies the raw primal psychic potentials capable of evocation through images.

The zodiac itself is a circular image, the word in Greek meaning "circle of animals." While the zodiac as a formal coordinate system originated with the Babylonians around the seventh-century BCE, it has been hypothesized that the 17,000 year old paintings of animals in the Lascaux caves represent the constellations in what could be described as a kind of proto-zodiac.[13] This proto-zodiac hypothesis is difficult to prove, however it is generally accepted that at the very least the cave paintings represent a form of "sympathetic magic," wherein painting the animal becomes a kind of spiritual communion between the souls of the animal and the painter.

Evidence suggests that early modern humans existed in an undifferentiated state of consciousness and lived by a worldview which anthropologists call animism. For these people, there was no separation between the spiritual and material worlds, and so animals were naturally seen to have souls too. In fact, in an animistic consciousness, everything has a soul—including rivers, mountains, valleys and plants, minerals, etc. So, the ritual act of creating the image of an animal is practiced to enable the artist to invoke the sympathy of the animal's soul—either to gain its sacrifice in the hunt or to take on some of its attributes and power. Perhaps this sympathetic magic is also partly what was intended in the creation of the zodiac, and helps explain its continued popularity. After all, who has not occasionally wanted to roar like a lion in the face of life's trials?

It is one thing to wistfully wish for the presence of one's inner lion, or even to accidentally stumble upon and arouse it, but it is quite another thing altogether for someone to consciously and deliberately summon such a presence. Outside of a children's story, many modern people might find the concept laughable. And yet, ironically, we can regularly see humans engaging in behaviors that make those of a wild lion seem almost tame. Is it possible that suppression or repression of our more primordial urges has only fated us to become possessed by them, forcing them to reveal themselves to us in a more perverted form? Perhaps the hypnotic allure of modern motion pictures, television, and now "selfies" can be explained as a natural psychic hunger to balance today's scientific/rational emphasis with a return to the more primordial image.

Philosopher Jean Gebser theorized that humanity has transitioned through several modes or structures of consciousness.[14] The problem with what he calls the mental structure that humanity is transitioning away from is that it seeks to deny the other structures with its claim that humans should be exclusively rational. However the structure that we are transitioning toward is integral, and carries the need to "make present" *all* the various structures of awareness. When all structures are recognized and accepted this enables a person to see and live *through* the various structures simultaneously, rather than be subjected to, possessed or "lived by" one of them. Perhaps this tells us that an openness to and understanding of an animistic perspective or worldview may help (or at least begin) to provide or reconnect us with conscious access to the ancient instinctive resources shared by all human beings, and might also help prevent these same instincts from taking over our lives through unconscious animalistic behaviors.

Although it is quite common for people today to think of god in terms of a trinity, it seems seldom that modern people dare to think of themselves, their lives or their world in these same terms. Yet, some familiar with more ancient forms of awareness know that animistic cultures have long considered there to be three worlds. In many ancient religious systems, there were three cosmic levels: not only heaven and earth, but an underworld as well. Rather than simply the nightmarish vision of hell imagined by Christianity, the underworld was seen by many cultures as a place of natural riches and ancestral wisdom. In this worldview, the *axis mundi*, the vertical feature of the cosmos, was seen to be at the center of the world and served to link together all three cosmic levels. This axis could be represented by various symbols such as a mountain, tree or ladder.[15] From an animistic perspective, all three of these worlds are not only connected, but also accessible to, and indeed part of every human being.

From an animistic perspective, because there is no separation between the material and spiritual world, we can also conceive of three "selves" with which to navigate these three worlds. For the animist, what most people think of as their entire identity, the every-day awareness of the conscious ego, or what Kahuna shamans call the "talking self," is actually far from the totality of being.[16] Our "higher" spiritual self has access to transcendent spiritual wisdom, and our basic self accesses our "lower" animal/visceral intelligence, the instincts and inherited tribal wisdom that have kept us alive as a species for many millennia. From this perspective, the admonition to "know thyself" takes on new complexity. There is a need to comprehend, understand, and harmonize all three essential aspects of being human and to bring our three "selves" into alignment and integrated partnership.

In this way, the triple-alignments of Mercury, occurring in the same degrees and also linking the above and below, can become a kind of *axis mundi*—a sacred linkage, connecting the three worlds and three selves. Remember, from a visual standpoint, the image of Mercury's retrograde journey is that of a disappearing act: from above to below, and back to above. Images often tell a story. The visual transformation process that Mercury performs every four months, of disappearing in the west and later re-appearing in the east, also happens to the other planets at various intervals and was mythologized by the Babylonians and Egyptians as the journeys of various gods through the underworld.

Of all the Greek gods, only Hermes was able to fully traverse the *axis mundi* and visit the heights of Mount Olympus as well as the depths of Hades. Sky astrologers can use these visual and mythical perspectives to better understand Mercury retrograde.[17] After Mercury passes evening elongation, his highest appearance above the horizon in the west at dusk, he is in the process of slowing down and descending. Later he turns retrograde and becomes invisible, disappearing in the west. After making the invisible inferior (below) conjunction with the sun, Mercury re-appears in the east and then makes his highest appearance above the eastern horizon at morning elongation. Visually, Mercury is "switching skies," appearing in the same degrees three times: first as evening star, then becoming invisible and making the inferior conjunction, and finally crossing for the third time as morning star.

If we process this visual image of Mercury's transformative journey as a story, and we follow along, it can be seen as a signal for us to "switch selves." Human beings are hard-wired with an instinctive response to the 24-hour cycle of light/dark known as the circadian rhythm. When the sun disappears in the west it is a signal that we need to take shelter and prepare for sleep and dreamtime. Similarly, when a planet dips below the threshold of visibility, it is signaling for us to take part in a concurrent descent into our basic selves, in order to reinvigorate our awareness with its special gifts. During what I call his "backward trickster medicine dance," Mercury takes on his most complex mythic role: the psychopomp, the guide of souls through the underworld, that is, the lower world of the basic self, animal powers, and instinctive wisdom.

Between the worlds, thresholds exist. For the Greeks, Hermes was the god of these liminal places. This is why Hermes frequently shows up in myths to give aid to heroes (for example, Priam and Odysseus in Homer's *Odyssey*) who journey into the unknown, as well as to escort departed souls along their journey into the final unknown of the afterlife (for example, the "suitors" in Homer's *Odyssey*). This is also why travelers would place a stone onto a herma, a heap of stones found at transitional places like crossroads and graves: to honor and seek the protection of Hermes, the god of the doors to the unknown. Thus, it is only natural that the Mercury retrograde experience may sometimes include messages or messengers of transition.

If we are not conscious of the deeper purpose of these transitions, then we can seemingly become their "victims." The underworld journey is not rational, and so any expectations of a strictly "normal" experience from our rational consciousness are likely to meet with disappointment. However, this does not mean we should simply superstitiously avoid certain activities during Mercury's retrograde. Rather, we need to try to learn or remember a different kind of dance altogether! First we must honor Mercury's trickster medicine by adjusting our own expectations and awareness. Only then we can begin to enter into correspondence, learn and remember how to dance with it.

While perhaps sometimes unwelcome to a strictly rational consciousness, properly understood these messages or messengers of transition can be seen as an invitation to "switch worlds" because mythically they serve the role of threshold guardian: the sentry at the door to other worlds.[18] These sentries serve to bar entry to those who are not ready to accept the magic of transformation in their lives. Those who play the "blame game" by making Mercury into a scapegoat and themselves into victims are refusing the magic of transformation because it does not fit into their limited rational understanding. The I Ching tells us that when faced with an obstruction, the common person looks to blame others and fate, but the noble person looks within. Like the common person who cannot imagine that there may be some deeper, unknown and mysterious meaning and purpose behind an obstruction, Gebser's mental structure of consciousness does not want us to entertain the viewpoints of the other structures. The mental structure wants rationality to be seen as the complete reality, in and of itself. But it is only when we can break free of the stranglehold of rationality that we can hope to see reality more as it truly is—infinite. Or, to further invoke William Blake, only when the "doors of perception" are cleansed of an oppressive uni-polar viewpoint of a singular truth, can we begin to see reality as multi-dimensional and multivalent. We must periodically traverse the passages between the rational and non-rational worlds if these passages are not to become the closed doors of a prison.

Somewhat paradoxically, it is only through a descent—"lowering" our awareness to include the non-rational levels of the mythical and magical—that we can ascend and "raise" awareness to the integral level, and learn to see across or through all the structures of consciousness simultaneously. There are good reasons why Mercury should be considered the appropriate god to teach us about this descent into the primordial. First, there is evidence that "the god of stone-heaps existed in Greece before the coming of the Greeks."[19] Second, Mercury was often seen as a tutelary or protective divinity, and was not only connected to domestic animals as god of herdsmen but also connected to the wilder beasts which still beset the traveler or wayfarer in those days. Therefore, as Mercury descends from evening elongation in preparation for his invisible journey we can learn to practice our own descent into the unseen mysteries of sympathetic magic and myth, thus creating and maintaining a conscious identity for our basic self through exploring our own personal animistic mythology.

To someone with an animistic worldview, animal helpers are a tremendous source of personal power and protection. Often the shaman will initiate healing by re-establishing the bond between a person and their power animal. Everyone has an animal protector or totem, if they are unaware of it. Whether you refer to the western tropical zodiac, the Chinese zodiac, the Mayan day signs, or simply your life experiences, there will certainly be an animal or two that you feel connected to, and perhaps a certain pride in because you exhibit similar character traits.[20] It may even be an animal that lives in the same element in which Mercury was retrograde during the year you were born.[21]

For instance, I was born in 1968 when Mercury retrogrades were happening in air signs, and so my personal Mercury elemental year is air. In 1989, I unexpectedly received a spiritual awakening through a series of powerful encounters with several air animal totems (flying creatures)—first in my dreams, then through immersion in nature and further investigation of spiritual topics, and finally while on vision quest. It was this initial awakening that put me on the path to becoming an astrologer. After being initiated into astrology by a dream in 1993, I found my way into the astrology community in 1995 through Project Hindsight—a project to translate and publish ancient Hellenistic era astrology manuscripts. By 1989 the Mercury retrogrades had returned to air signs for the first time since I became an adult (the third return after my birth), and in 1995 when I found my way into the astrology community the Mercury retrogrades had again returned to air signs for the fourth time since I was born. In fact, this is a regular cycle which you can observe: given that Mercury only retrogrades in the signs of a particular element about every six to seven years, it follows that approximately every six to seven years, the mercury elemental year will return to your birth element.[22] In 1995, by descending into a primordial experience of the air element, I was able to find deep within myself a connection to a more integrated intellectual tradition that honors the magical and mythical as well as the rational. I was then able to ascend and return to my intellectual life in a more authentic fashion. It took me seven years to complete this journey, largely because I did not have any idea what I was looking for, except a connection to something more magical than academia seemed to offer me at the time.

This six to seven year cycle is what we might call the "long form" of Mercury's transformative dance, and we will explore it much further in chapter five. The "short form" of this transformative dance is contained in the approximately forty days between evening elongation and morning elongation. Mercury's inferior conjunction happens right in the middle of this roughly forty-day period.[23] Forty days is a very sacred number in many myths, and so we can use this smaller period for more targeted searches and as a sacred and safe container for a journey with a more clearly defined start and end date.

To keep the passages between worlds open, and our basic self and non-rational worlds integrated, we should dedicate at least one of these three periods each year to taking a conscious descent into the worlds of magic and myth. This may be as simple as taking time to immerse ourselves in nature without timetables, connecting physically with the elements, and feeling the heartbeat of nature in a personal *Walden*.[24] Or perhaps your "road less traveled" might be found by exploring myths or even inventing a new personal mythology. It can even become as wonderfully complex as your own *Red Book*, perhaps by keeping a personal journal or sketchbook which chronicles your experience of connecting with various animal or plant energies.[25] Anything that introduces an element of the sacred, magical or mythical regularly into your life during these forty days has the potential to become transformative. I offer a few basic exercises below as starting points for your journey.

Spiritual Practice: Gathering and Remembering Medicine

Around the time of evening elongation, seek to slow down and make time to enter a liminal place, free of linear schedules and expectations. Spend time in or near nature, especially liminal areas such as borders or edges where two kinds of habitat meet or overlap. Pay special attention to the threshold hours of dawn and dusk as well. See a sunrise or sunset, and witness the changing of worlds. Become aware of and identify any animal or plant medicine that enters your awareness, whether in waking life, daydreams, or sleeping dreams. Spend time in or around their element. Notice carefully what these beings do, what feelings are aroused in you through their presence and contemplate the wisdom and lessons in each. Seek to establish or reestablish connection with your animal or plant totem(s) through ritual images and objects, such as artwork, songs, personal stories, or mementos and create a medicine bag, altar, journal, or other sacred vessel to consecrate and contain this medicine. Remember to honor and thank your totems with offerings, prayer, contemplation, and practice embodiment of their sacred medicine.

Spiritual Practice: Engaging the Living Myth

Around the time of evening elongation seek to slow down and enter a liminal place as if you are waiting for a special visitor to arrive. Begin to see any challenging people or events that emerge in your life at these times as special messengers, welcome them with grace, and let them know they are accepted and appreciated. Remind yourself that these are threshold guardians and by showing them respect you can gain entry into the special worlds, which contain the gift of wholeness.

Now, enter into correspondence. Spend some time chatting with these people or events, even if the conversation is "imaginary." Ask them what they have seen or heard. Appeal to their knowledge and expertise and ask for pointers, tips or insider info. Be sure to thank and honor them and ask if you can be of assistance to them, knowing that you are communing with the spirit of Hermes through them.

After these discussions, take some time to return to these stories and examine them with your rational mind as if it were a dream or myth, looking for the allegorical and metaphorical spiritual lessons in the story, and imagining what the next parts of the tale might look like once the heroine understands what is being asked of her.

Spiritual Practice: Fire Meditation

The archaic structure of consciousness can be associated with the pre sapiens species of humans such as australopithecus and homo erectus.[26] Even these pre-modern humans were able to manipulate tools, which symbolizes the beginnings of mastery of the earth element. However, other animals such as crows and ravens are also capable of both making and manipulating tools. However, it is the controlled use of fire by these early pre-modern humans that truly sets them, and eventually us, apart. Our relationship with fire then becomes the door into the archaic structure of consciousness, before we were aware of any separation from nature. As mammals, with the ability to maintain a temperature of 98.6 degrees Fahrenheit, our bodies are quite literally furnaces. The metabolic processes that break down food are like the flames of a fire consuming twigs and sticks, turning them into ash. We constantly radiate this inner fire through our breath, skin, and excrement. And so we must feed the fires of this inner furnace that keeps our bodies moving. When we eat we are taking in sacred fire, the energy of the sun, stored first in plants and then in other animals. Even the sun, the source of most of our energy, must consume energy to create iron and all of the heavier elements. So we can see that life is simply the constantly flowing process of energy exchange throughout the universe. This living fire represents the deepest part of who we are, and yet it does not belong to us at all, but rather simply flows through us—as it flows through all life.

The Serpentine Fire Meditation

Around the time of evening elongation, create a quiet time and space to feed your soul with the everlasting fire of life. At night, or indoors, light a fire in a safe place such as an outdoor fire-pit or indoor fireplace. Even the flame of a small candle will do. Seat yourself comfortably in front of this flame and allow your mind to become quiet as you gaze into it. Notice how the flame flows upward while constantly undulating at its base, like a snake that has been charmed. Feel this movement in your body, slowly undulate your torso and feel the sa-

cred fire flowing through your body, rising from the base of your spine and culminating at the top of your head. Imagine now that there are two serpents dancing together in this sacred flame, slowly swinging back and forth, writhing together until they are standing erect intertwined, like the dual serpents of Mercury's caduceus.

Now direct your attention outward again, and visualize this same fire in everything around you. Realize that every creature, every leaf, every breath of wind, and every grain of sand quivers and pulses with the same sacred fire of life that flows through you. Feel this connection in your heart and soul and think of all the many beings who have shared this everlasting fire with you. Give thanks to them for their gift.

Now, enter into communion with the sacred spirit of fire by imagining your own inner flame flowing from the base of your spine, up and out from the top of your head and toward the fires of all creation. Imagine a central fire, like a great sun sitting inside an ancestral well, Your fire stream, like all the other fires, flows into this central well of fire, commingling for a moment and then returning to you as a stream of pure white light to enter at the base of your spine. Now imagine this stream of fiery white light circulating, flowing through you, from the base of your spine and up through your body, out the top of your head and back to the great ancestral fire of creation, where all fires join, and then back to you once again. Feel this flowing stream of white fire permeate your being and imagine it taking the shape of a lemniscate or figure eight as it flows through you, through the worlds and the great ancestral fire. Relax and allow this circulation of fire to flow of its own accord, replenishing and renewing every cell in your body.

Emblem 21 from Michael Maier's *Atalanta Fugiens*, 1617.

Chapter IV
The Alchemy of Transformation

> *Both the imagination of mercury and the will of sulphur must be called... When the imagination creates an image and the will directs and uses that image, then marvelous magical effects can be obtained.*
>
> —A NINETEENTH-CENTURY GOLDEN DAWN ALCHEMIST[27]

IN THE PREVIOUS chapter we explored the necessity and transformative power of a conscious descent into the primordial non-rational parts of our nature. It is only by first embracing both sides of a polarity that we have any hope of reconciling them. This way, rather than renouncing our primal nature or withdrawing into the ivory tower of intellect, we can instead seek to refine both sides by drawing them together. This is one of the foundational tenets of alchemy.

In the facing image, we see the alchemist drawing our attention to the relationship between the world contained in the smaller circle (material world), and the world contained in the larger circle (spirit world)—that is, between the microcosmic and the macrocosmic. By attending to both worlds, and seeking to understand the principles that apply to both, the alchemist nurtures an integral consciousness perfected by the great work. By recognizing, understanding, and working with both the spiritual *and* material dimensions, and seeking to bring them together as a whole, we may learn to come into correspondence and union with the innate wholeness of life itself.

Just as the alchemist in the preceding figure continually works at understanding the dualities of spirit/matter, above/below and rational/non-rational, in the same way, we can continually work to understand the relationship between the times when Mercury is direct *and* the times when Mercury is retrograde. In fact, observe that it is a triangle which the alchemist uses to connect these two worlds, just as Mercury traces a pair of interlocking triangles in the sky each year. Thus, recall that the triangle traced by Mercury's inferior conjunctions derives its transformative power from the

Tria Prima emblem from the section on Valentine's *Twelve Keys* in Lucas Jennis' *Musæum Hermeticum*, 1625.

above/below correspondence in the triple-alignments which happen in the same degrees: first crossed in direct motion (evening elongation), then in retrograde motion (inferior conjunction), and then again in direct motion (morning elongation). Because each of these three events has an implied relationship with the others, they form what can be called a differentiated unity. Essentially, by working with any of these alignments, we are invoking a magical formula, one known by writers as "the rule of three" and by artists as the "rule of thirds." These triads contain the smallest amount of information to create a pattern, leading the prose and art created with them to be more memorable and effective.

In fact, a structure involving three principles is also a fundamental part of the Hermetic art of alchemy. The spirit/matter dichotomy we used before to grasp the opposing directions of Mercury's movement (direct/retrograde and above/below) was understood in alchemy as the pair of opposing principles: Sulphur (dry/masculine) and Mercury (moist/feminine). When these two opposing principles are joined, we get Salt (body). Thus, we have three essential principles, the *tria prima* or three primes: Sulphur, Mercury, and Salt, also known as *anima* (soul), *spiritus* (spirit) and *corpus* (body).

In the image facing, the flask or alchemical vessel of transformation contains a dragon inscribed within a triangle and surrounded by three serpents. Inside the triangle, inscribed upon the dragon's body are two circles representing both its winged-ethereal (volatile/spiritual) aspect and its earthly-bodily (fixed/material) aspect. Outside the triangle, the three serpents move in a circle around the dragon, representing the encompassing nature and transformative powers of the tria prima.

The serpent that moves downward can be seen as salt and represents a contractive, restricting, crystallizing tendency similar to the mental structure of consciousness and the rational discriminative powers of the every-day "talking self" or ego. The serpent that moves upward can be seen as sulphur, representing an expansive, radiating, dissolving tendency, similar to the non-rational structures of consciousness and the imaginative powers of the "lower" or basic self. The serpent at the top can be seen as mercury which, as an alchemical principle, represents a flowing, interweaving force of dynamic equilibrium that always seeks to balance any pair of opposing forces in any outer phenomenon or inner spiritual experience. Mercury can be seen as similar to the integral structure of consciousness and the transcendent powers of the "higher" spiritual self.

As fundamental principles, the *tria prima* are omnipresent. The Swiss alchemist Paracelsus reasoned that the four elements appear in bodies by way of these three principles, using the analogy of three parts of wood that are revealed by fire. The heat-giving flames describe flammability (sulphur), while the smoke describes volatility (mercury). Mercury includes the cohesive principle, so when it leaves in smoke, the wood falls apart, and the remnant ash describes solidity (salt).

It is not a coincidence that the author of many of the Hermetic texts was [known as] Hermes Trismegistus or "thrice-greatest." In addition to his triple alignments, many other things about Mercury come in threes. For instance, Mercury's glyph is the only one of the classical planets to contain all three elements: the "crescent of soul" above the "circle of spirit" above the "cross of matter."

Thus, Mercury's glyph is a symbol of wholeness—the bringing together and reconciliation of opposing principles in order to exist on the material plane in physical form. The three elements of this glyph symbolize the potential alignment between the *three selves*—i.e., "lower" basic self (instinct/soul), "higher" self (intellect/spirit), and every-day "talking" self (material self or ego/body). Psycho-spiritually, these three principles are also a differentiated unity; we must find a way to integrate all these qualities in ourselves and our lives for effective and successful living.

This premise also works in reverse. We can also use the *tria prima* to first *dis*-integrate a problem. The basic formula in alchemy is *solve et coagula* or "break apart and put back together." So when a problem arises, assuming we are able to keep from playing the "blame game," we can do much more than simply try to address the material manifestation (body) of the problem. We can break it apart into its three primary principles.

The basic idea is actually very much like modern depth psychology. Treating the symptom or dealing only with the material manifestation (body) of the problem is like pulling weeds without ever getting at the root. You will just have to go back and do it again and again. By going deeper and getting to the other fundamental principles (soul and spirit), we have a much better chance of actually curing the condition rather than simply temporarily alleviating the material symptoms. However, as we've noted before, to properly perceive these other principles, we have to move our attention from the physical world to the spiritual world. Only by turning within and undertaking a conscious descent can we achieve the breakdown (*solve*) that will ultimately result in a breakthrough and restoration of wholeness (*coagula*). The good news is that the alchemy of three provides us with a map for this journey on the road less traveled.

Mythologist Joseph Campbell divided what he calls the monomyth into three basic phases: separation, initiation, and return.[28] Just as Campbell's structure was inspired by an earlier work,[29] this threefold division of the myth/cycle mirrors a profoundly archetypal process of transformation which is ubiquitous and can be seen not only in the three classic phases of alchemy (*nigredo, albedo, rubedo*), but also in such varied examples as the triple way of mysticism (purgation, illumination, union),[30] the way of the shaman (training, initiation, re-birth), and even in the phases of human birth (contractions, delivery, reunion with mother). Psychiatrist Stanislav Grof has theorized that the processes of human birth are so profound that they form permanent structures or matrices in the human psyche.[31] Thus whenever we experience significant change (or re-birth) in our lives, these basic structures are spontaneously reactivated.

Nigredo, albedo and *rubedo*—this is the hidden order within the apparent chaos of Mercury retrogrades and in fact the pattern we can expect whenever we journey from the known into the unknown. They are like three gates or doors we pass through on any journey in search of wholeness. These three gates represent passages to three worlds, and three primary aspects of our psyche, the *tria prima* of self. Now, let us examine each of these stages in depth, so that we may be more familiar with the archetypal map, and more likely to find our way, no matter into what rabbit holes we may fall.

EVENING ELONGATION, SEPARATION, NIGREDO

After evening elongation, Mercury appears lower each evening in the west at dusk, before turning retrograde and finally disappearing altogether. In the previous chapter, we talked about seeing this as a signal to "switch selves" and descend into the non-rational, primordial worlds of magic and myth. The purpose of starting in such a fashion is to first re-animate the individual and their life, to get them in touch with their soul and will, in preparation for the difficult work of transformation. We can think of it like stocking up on wood and food before a long winter storm, or filling the gas tank before a road trip. We do not want to run out of fuel in the middle of the desert.

Once we have recovered enough anima, chi, life-force or soul, then we can more safely enter the fires of purgation, the first stage of the transformative dance. Nigredo means blackening and refers to the color arrived at when initially combining the base metals, as well as the accompanying psycho-spiritual states and imagery. The nigredo stage is often symbolized in alchemy by images such as the black crow, skulls and skeletons, death and decay, and purification by fire. However it is important to realize these are not simply nihilistic images without purpose. Rather, the overwhelming darkness and suffering of the nigredo stage also contains the small hidden seeds of light and healing.

Often the Greek myth of Coronis is used to illustrate this point. Coronis was pregnant with Asclepius. The father Apollo left his animal familiar, a white crow, to guard Coronis while he was away. Coronis had an affair and when the crow delivered the news to Apollo, he was so enraged that the crow did not somehow intercede, the force of his fury was flung as a curse which scorched the bird, turning it forever black. Apollo then sent his sister Artemis to kill Coronis. As Coronis burned on her funeral pyre, Apollo came to his senses and had his little brother Hermes rescue the baby Asclepius. Hermes then gave the baby to the wise centaur Chiron to raise, and Asclepius went on to become a great healer. Thus we can see the seed of new life and healing dwells hidden within tragedy.

There is another related myth involving Hermes with the breaking of social taboos. Hermes is actually said to be the first of the gods to be stained with death. In this tale, Zeus sent Hermes to free his lover Io, whom Hera had guarded by the monster, or primordial giant, Argus. Hermes put Argus to sleep with song and story and then killed him, and so was put on trial. The gods were aggrieved by this, but since Hermes was acting as the agent of Zeus, they were nonetheless forced to acquit. In a seeming act of symbolic stoning, the other gods cast their votes by throwing their voting pebbles right at Hermes, wherein a heap of stones grew at his feet. This myth not only explains the connection of Hermes with the ubiquitous primordial rock cairns known as *herma,* but perhaps also deepens the sacrificial role of Hermes, who in the Homeric hymn becomes the god of animal sacrifice in the first days after he is born.[32]

The psychological lessons of these myths seem to be that psycho-spiritual re-birth is necessarily preceded by ego-death, and that sacrifice is often a necessary precursor or catalyst to growth. Perhaps this is why so many rituals begin with purification, which can be physically, emotionally, and/or spiritually painful. The purification of nigredo can be achieved through physical processes like sweating or fasting and psycho-spiritual processes like solitude or therapy, which force us to "face the dead parts," that is, the outworn or destructive parts of our personality and lives.

Just as every inferior conjunction of Mercury is accompanied by an evening and morning elongation in the same degrees, every ego has what Jungians call a shadow and a persona, a tendency to conceal and reveal itself. When un-integrated, the persona and shadow complexes can operate in a Jekyll-Hyde fashion. The process of nigredo involves facing the need for deep inner changes to these psychic structures in order to facilitate their later growth and integration. In the process of personality development we consciously develop traits to meet the demands and expectations of society in order to satisfy our own social aims and aspirations. Jung called this revealed, socially acceptable veneer of the ego, the persona. The shadow is the concealed unconscious counterpart to the persona. Just as Mercury runs counter to normal direction during its retrograde, the shadow contains features of our nature that run counter to the customs and moral conventions of our society. Psychologically, the shadow functions as a door or bridge to the personal unconscious and contains personal characteristics of which we are ignorant, afraid, or ashamed.[33]

Though it is ultimately a source of growth and individuation, the ego's defensiveness causes both individuals and groups to tend to "cast" their shadow, and ascribe their own dark traits to others. Thus, we often meet the shadow through "the other"—other people, groups, or events involving others. Before integration, this often results in the "blame game," where the shadow takes the form of people we label somehow "bad" or unacceptable, while we assume the role of "victim." As we begin to integrate shadow material, this can also involve the process of "tough love," wherein a supportive person whom we love or trust challenges us to put negative shadow behaviors behind us. At higher levels of integration, there is even what is called the "golden shadow," which consists of parts of ourselves that are ultimately positive, but are so powerful that we are afraid to embrace them. Going against the grain of society, even when ultimately helpful and productive, is scary and can often even be dangerous.

The purpose of the nigredo process is to purge the body, the psyche, and the life of extraneous parts or "dead weight" in order to free up energy for future growth. Psychologically the persona and shadow need to be periodically purged, so that we can later update each of them to form a more authentic and integral team. We can understand and achieve this purgation by using the *solve* or "break apart" element of the alchemical formula. For instance, if the persona is what we consciously identify with as the socially acceptable "me" (the revealed self) and the shadow is that we have labeled "not me" (the concealed self) then by simply applying the filter of time we can break these apart into two categories: that which is "*no longer me*" and that which is "*not yet me.*"

At dusk, when the sun is setting in the west, the shadow is behind us to the east, and thus symbolically represents everything that we have already experienced that day. We have also seen that evening elongation also signals a time of impending descent. Therefore, Mercury at evening elongation is a time to "face the dead parts" and deal with what is "no longer me." We must identify and cast aside the concealed negative behaviors of the shadow that limit us. Jung identified Freud's notion of the id with the shadow, that is, in its negative form the shadow is primarily interested in selfishly attaining pleasure and power. The negative shadow behaviors of an over-functioning id might also include "normal" pleasurable activities or socially acceptable vices that are becoming unhealthy addictions. Also, though always somewhat concealed or denied, any overly aggressive power-seeking behaviors or abusive misuses of power are manifestations of the shadow. Conversely, an underfunctioning id or basic self may manifest as the passive-aggressive failure to claim our healthy power by playing the "blame game" by making ourselves into a victim and someone else into a scapegoat. This also falls in the realm of shadow behavior, and we must periodically cast any of these shadow manifestations into the fires of the nigredo, in order to continue growing.

We must also periodically identify and cast aside the dead parts of the persona—the people pleasing aspects of our personality that are no longer necessary because we have outgrown them or because they are hindering our further development. Often, it is only by formally recognizing and then sacrificing our previous growth that we can make room for further growth in our lives. A gardener must clear away and compost the detritus, the dead or decaying stalks, vines, and leaves at the close of each season. So too must we periodically recognize, honor, and ultimately sacrifice and replace the supportive structures in our lives that have allowed us to become who we are. Sometimes it is only by letting go of the past and clearing room in our lives and our psyches, that we can become aware of—and eventually consecrate—the "not yet me," the future within us, that which is seeking to be born.

Spiritual Practice: Purging the Salt
The Potlatch or Give-away Ceremony

In the journey of life we are constantly accumulating "stuff." It is great to have possessions which carry deep meaning or serve a purpose, but the stuff we continue to carry after it has lost its meaning or usefulness to us is just "dead weight."

After evening elongation, as Mercury begins his descent, search through your possessions and find items that you no longer identify with or use. Give away one or a few of these items to a neighbor, friend, or stranger with no expectations for reciprocation. If you have many items, give them to a charity or family. Or you can have a garage sale where everything is free.

Fire: Artwork, exercise, sports, or outdoor equipment and apparel
Water: Items with emotional attachments or old family items (children's clothes, etc.)
Air: Books, movies, or music
Earth: Tools, clothing, kitchen or other household items

Now go back and search through your possessions for the items which are most dear to you, with the most meaning or usefulness. Find a way to create or purchase a duplicate, copy or a reasonable facsimile of one of these items and watch for an opportunity to give it away to someone whom you think could use its special medicine in their lives or who needs to be acknowledged for their spiritual growth.

Now go back again to those items which are most dear to you, with the most meaning or usefulness. Look especially for the ones that mark a previous rite of passage such as an achievement or an especially holy moment. Consider what it would feel like to practice the give-away with such an irreplaceable item. Is there someone in your life whom you could recognize with such a gift, passing on your own memento to mark a sacred moment for another? How might that catalyze your own need for another rite of passage, to reach another mountain top? It will not be every cycle or perhaps even every seven years that we need to literally practice this level of give-away, but it is always healthy to contemplate.

Spiritual Practice: Purging the Sulphur
The Fasting Ceremony

Many spiritual traditions include the fasting ceremony, particularly around special holy days. Many people think of fasting primarily as a bodily cleanse, which it certainly is. But the deeper purpose of fasting is to purge the will, the spiritual correlate to the inner fire or furnace. By summoning the willpower to release our attachment to a pleasurable item or activity, we are reminding ourselves of our ability to endure hardship with grace, discipline, and faith. Through fasting we can also release the illusions of powerlessness and victim roles and re-claim our co-creative power.

Beginning around evening elongation, and afterward as Mercury descends, watch for things (people or events) that challenge your attachments, especially if a sense of loss or feelings of entitlement arises. This is a message from Hermes to consider fasting from this item or activity, in order to form a more healthy relationship with it (note: only literally undertake a fast if doing so would not represent a serious health risk—i.e., you do not want to suddenly fast from all food for too long or suddenly stop taking medication without consulting a physician). If completely fasting from the item or activity seems too much, perhaps you could just cut back on your involvement a little bit or simply monitor yourself for the feelings of loss and entitlement and fast from acting on these feelings—practice observing the feelings without reacting to them.

After evening elongation, if the spirit of Hermes does not present a challenging person or event which could indicate something to practice awareness of attachments around, consider honoring him anyway by choosing to voluntarily fast from something for a few days, and if that goes well, perhaps continue the fast for a week, or more. Simple things to practice breaking attachments to are "luxury items" like sugar or sweets, alcohol, or adornments. We can also practice fasting from behaviors, like critical comments (if you do not have anything nice to say, say nothing at all) or judgmental thoughts (practice observing them and letting them go without following them or perhaps balancing them with supportive or loving thoughts).

Spiritual Practice: Purging the Mercury

The mercury principle is the most elusive of the *tria prima*. To deal with mercury directly, we need a way to become the self-reflective observer of our own consciousness. In order to become the observer, we first need to create a safe space where we are free to express ourselves as naturally and authentically as possible.

Particularly, as the "concealed self," shadow behaviors and dynamics can be especially hard to get at and tenacious because of a natural defensiveness. We fear that if we reveal our unacceptable shadow it will bring the socially acceptable persona crumbling to the ground, and of course this fear has a very real basis in human history. However, it takes a lot of psychic energy to keep up these defenses. Over time, the stress of inhibition can take its toll on the body. This is why the basic idea behind any kind of psycho-spiritual therapy, whether it be in the Catholic confessional, groups like alcoholics anonymous or a therapist's office, is to create a sacred environment in which people can safely reveal themselves. While seemingly simple, this opening up, expressing and revealing of ourselves in safety can have profound ripple effects throughout the system—not just on the psychological functioning of an individual, but even on physiology.[34] In fact, there is a virtual cornucopia of research over the last twenty years, which supports the idea that the so-called "talking cure" of psychotherapy can also be transformed into a "writing cure." Journaling, keeping a diary or expressive writing has many advantages, and it is much easier to get started than you may think.[35]

Contemplation, Meditation, Journaling, Art
Group or Individual Therapy

After evening elongation, as Mercury begins his descent, create some time and space for private expression. In my experience it is possible to create your own "Red Book" adventure by learning to express yourself via multiple media. Even if you do not feel comfortable being artistic you can simply search the web for images and music to accompany your latest thoughts and feelings, bring them all together around a central theme and keep these posts in a private journal or blog. By expressing the shadow and bringing it out into the light we can create a more conscious, intimate and satisfying relationship with the three most important people in our lives—me, myself and I.

Fire: *Identify ways you are spending your energy in a grudging manner, because you feel obligated*
Water: *Identify negative feelings you are having that have their roots in past situations*
Air: *Identify negative thought patterns, especially those that belong to old belief systems*
Earth: *Identify areas of your life that feel too structured, rigid, controlled, or stagnant, especially those rules, boundaries, or limitations that are being upheld out of fear of reprisal rather than healthy respect for tradition*

INFERIOR CONJUNCTION, INITIATION, ALBEDO

The second time Mercury passes the (eventually) thrice-crossed degrees, he is in his underworld incarnation—hidden below the horizon and invisible. Astronomically, at the inferior conjunction Mercury is at the point closest to earth for the entire synodic cycle, thus accomplishing a linear alignment of earth, Mercury, and sun. As Mercury passes between the earth and sun spatially, he can be seen to take on a medial role, serving as a bridge between the limited material experience of earth and the enormous unlimited potentials of the sun as a star—the creative engine of the universe. Psycho-spiritually, this represents an opportunity for a similar alignment and mediation between the *tria prima* of self. We can now bring into harmonious alignment and achieve integration of our "three selves," so that both the animal/instinctual knowledge of the basic self and the higher wisdom of the spiritual self can inform the everyday awareness of the talking self or conscious ego. Classically, the exact conjunction was known as *cazimi*, meaning "in the heart of the sun," and it symbolized a meeting in the throne room of the King and Queen. Conjunction is the time when the solar seed of the above, the spiritual will of the inner sovereign or supreme leader (our higher self), is impressed upon the receptive fertile matter of the below. This is part of the *heiros gamos,* or sacred marriage of opposites: masculine and feminine, *yin* and *yang*, spirit and soul. The task for us during the heart of Mercury's retrograde is to become aware, and make conscious, our deepest internal opposite, to unearth the unconscious unlived parts within us which are seeking the light—like a seed germinating in the darkness. Will this seed of a new self germinate in the darkness of the cellar (our unconscious) and its potential be wasted? It is up to us to make sure this seed-potential somehow falls upon the fertile ground of our conscious lives.

In daring to visit our deep interior spaces, to speak to our inner sovereign, to listen for their dreams and, out of love, make them our own, we can invoke a profound magic. The conscious union of opposites was symbolized in the figure of the Rebis by the alchemists. *Rebis* comes from the Latin *res bina*, meaning dual or double matter. Like the ever-flowing and changing yin/yang, the principles of sulphur and mercury are sometimes difficult to distinguish from one another, so that the alchemists view them as two faces of one thing. The androgynous Rebis figure symbolizes a successful conjunction, combination or union of these two volatile principles.

Like the glyph for Mercury, the entire Rebis image is formed from three elements. The two-headed hermaphrodite holds the Masonic tools of study and proper alignment, indicative of both the awareness and ability to work on one-self (spiritus/spirit/mercury). It stands upon a dragon, thus representing the energy and spiritual will-power (anima/soul/sulphur) required for successful "taming" of the baser elements of one's unrefined character (shadow). The winged disk which supports both the dragon and hermaphrodite contains the triangle and the square, illustrating an understanding of the natural patterns in the archetypal emergence of all numbers and forms from the one thing, or *prima materia* (corpus/body/salt).

Psychologically, the rebis corresponds to the *anima/animus* complex, which is a deeper structure than the shadow, and represents an attitude that governs our relationship to the secret inner world of imagination, ideas, and dreams. As our internal opposite, the *anima/us* can initially reveal itself as the deepest feminine characteristics in a psychologically masculine person and the deepest masculine characteristics in a psychologically feminine person. Even in modern society, a man's deepest sensitivity and vulnerability and likewise a woman's deepest boldness and power are often

repressed for social reasons. However, whereas the shadow is concealed mostly out of social necessity, the *anima/us* is also hidden from us simply by its very nature and psychic depth, as a passage to the collective unconscious.

Beyond the complex lens of gender, the primary function of the *anima/us* is as a bridge or doorway to the collective unconscious or archetypal realms, which provide a ubiquitous source or wellspring of psychic energy and creative inspiration. Because of the universality of the collective unconscious, integration of the *anima/us* as our internal opposite serves to provide access to our deepest and broadest source of growth and individuality. It is through our opposite qualities we become aware of and correspondent with the archetypal realms, whereby we can re-form ourselves from beyond the "accident of birth" and the limitations of any particular sociocultural milieu.[36] Genuine contact and correspondence with the archetypal realms serves to channel the light of spirit.

The albedo (from latin *albus* meaning white) phase in alchemy is named after the pure white light that spontaneously appears after a successful transverse of the darkness of nigredo. Like the light of a new day, the color white is symbolic of purification, unity, and wholeness. It is as if, after descending through the dark cave of nigredo, we finally emerge into the light of the spirit world. The albedo is frequently symbolized by the white swan swimming gracefully on the surface, at the meeting place between the waters of soul below and the sky realm of spirit above. This new interface or relationship between the queen of soul and the king of spirit makes possible a deepening of the *heiros gamos*, or sacred marriage. While many modern people might confuse the words soul and spirit to mean generally the same thing, for the alchemists they are fundamentally opposing principles. Spirit lifts us up, whereas soul pulls us deeper into life. We can think of spirit as the inner engineer or architect of our lives. Spirit is masculine, *yang*, restless, constantly theorizing, striving for change, and making plans for betterment of the kingdom. The gifts of spirit are glimpses of the big picture and moments of freedom or personal accomplishment. Conversely, we can think of soul as the inner artist or storyteller. Soul is feminine, *yin*, wants to settle down, feel into, become one with and express the beauty of everything—as it is. The gifts of soul are gravitas, complexity, and rootedness.

Each and every one of us has something of an unconscious preference for the ways of the king of spirit versus the queen of soul. Yet, without both of them working in harmony we cannot become whole. To achieve this harmony we must search deep inside ourselves, and become conscious of the inner opposite to our outer expression. This begins a courtship where we watch and learn the ways and the wisdom of our personal path-less-traveled.

Many of the symbols alchemists use for the albedo phase involve the pairing of opposites: king and queen, sun and moon, stag and unicorn, two-headed lions or dragons sharing one body, birds chained or roped to earthbound animals, etc. The successful marriage or integration of two opposing principles produces a third thing, a child. This "child," or third consciousness, represents the natural fusion of soul and spirit. When the king of spirit's awareness of the future is wedded to the passion and vitality of the queen of soul then we can behold any situation with both the rational intelligence developed by the Greeks and the mystical 'intelligence of the heart" nurtured by the Egyptians. Like Gebser, consciousness pioneer and symbolist René Schwaller de Lubicz believed that the consciousness of humanity as a whole is evolving toward a new integral awareness, beyond the rational confines of science.[37] If this is true, then this third consciousness should be in the process of becoming increasingly accessible to more and more people.

The sudden emergence of the third consciousness during the whitening phase can produce a vision of the end product of the great work. This polar swing—out of the blackening into the blinding white light—can bring with it the appearance of seeds of the future. Through this catharsis, after an intense experience of being consumed in the fiery crucible of nigredo, we can glimpse the unveiling, however momentary, of a new possibility—a flickering light in our souls drawing us towards its promise.

Once we have looked within for our inner opposite, reached out and spoken to the inner king or queen, we must listen for their response and accept that whatever comes is a divine message, full of creative potential. If we can first do this, then around the time of conjunction we may be gifted with a glimpse of the future or even the ability to plant the newly conscious seed-potentials of our future. Far from being simply a time to avoid certain things, if approached from the proper perspective, the heart of the Mercury retrograde period can actually be an extremely powerful opportunity to practice electional astrology (that is, electing to begin important enterprises) and sacred magic. Despite pervasive superstition to the contrary, *if* the dream/vision has already been made clear *and* it is something that has either already been repeated twice or something where further repetition is desirable (i.e., where the rule of three is applicable), then one can absolutely choose to begin important undertakings during Mercury retrograde. I have personally signed contracts to sell important real estate holdings under Mercury retrograde and actually ended up getting more money than I asked for. In particular, the three days around the conjunction of Mercury with the sun are a special sacred window, and so can be a very powerful electional or magical moment. For instance, I posted the first episode of my astrology podcast on the day of a Mercury retrograde conjunction with the sun in Aquarius (air element) on February 6, 2008.[38] It has since been listed as one of the top astrology podcasts on iTunes for many years now, reaching number one twice. 2008 marked the sixth return of the Mercury elemental year of my birth (1968, air), and the conjunction of February 6 was also in the exact degree of the rising degree of my birth chart. A conjunction of Mercury with the sun in a particular degree of the zodiac can only happen once every seventy-nine years. It takes a lifetime to repeat the exact degree. So when these conjunctions occur in important degrees of a birth chart, they represent a very rare opportunity indeed: the potential for once-in-a-lifetime transformation.[39]

As Mercury passes between the sun and earth, the middle week of the retrograde period (especially the day before, day of, and day after Mercury's conjunction with the sun) represents a special time of alignment between the above and below. Allow yourself some time to dream of the future and its promise. On these sacred days you might seek guidance from an oracle such as the tarot, I Ching or horary astrology, embark upon a "vision quest," or work on a vision board for a new dream you want to realize. You may also wish to communicate your desires via prayer, by writing them down as affirmations, and by placing symbolic objects representing your prayers and dreams on your altar, in a hollow tree, or other sacred place.

Spiritual Practice: Illuminating the Salt
The Hour and Place of Power Ceremonies

Our sun is a star, and as such represents the creative engine of the universe—the source of all creative power. During the heart of Mercury retrograde, as Mercury aligns with the sun and earth, we have a chance to become more conscious of the currents of this cosmic power as it moves through our bodies and our lives.

The albedo phase represents a re-connection to and release of this cosmic power.

We live on a magnetic planet and the queen of soul provides us an internal divining rod with which we can learn to feel the currents of this power and learn to sense both its presence and its absence. A connection to power sends energy coursing through our bodies and our psyche and manifests as feelings of happiness, joy, excitement, boldness, courage, confidence, connection, unity and balance. Conversely, when our connection to power is blocked or lost we feel the opposite: sad, lethargic, hesitant, doubtful, alone, or unbalanced. During the middle of Mercury's retrograde period, make an effort to become more conscious of the currents of power, sensing and noting both its presence and absence in your body and your life.

Each of us has our own unique rhythm and adaptation to the daily cycle of light and dark. Around the time of inferior conjunction, become more conscious of the ebb and flow of your energies throughout the day and night. Notice the peaks and valleys of your energy throughout the day, realizing these may be different on the physical, emotional and spiritual levels. Notice also what activities, emotional currents and thought patterns are associated with feeling more powerful and which ones leave you feeling drained.

Once you have identified your "hour(s) of power," then it is time to give this information over to the king of spirit, who provides us with a natural sense of how to manage and arrange our various functions. Some people may want or need to hire the assistance of professionals to activate this function. This process involves trial and error and so requires patience and a methodical approach. Start by making small shifts in your schedule such as moving things by a half hour, and alternating between or beginning and ending on power activities. You might also want to make shifts in your diet, use feng shui to re-arrange the energy in your environment, or consult an acupuncturist or energy worker to help you move energy in your body. As more power becomes available, you will be able to risk making larger shifts until you have optimized your energies.

No matter how effective we are at managing our power, there are times where life will drain our batteries, and at these times we need to know how to re-charge. Or we might want to seek a boost before taking a risk or engaging in something important. Just like our bodies have energy centers and meridians, so too does the body of the earth. When spending time in nature, if we can quiet the mind then the queen of soul can help us to identify and locate places that naturally charge our batteries or serve to move and balance our energies. Unless there is a need for a radical energy shift, these need not be famous "vortexes" or healing springs. Usually there are sufficient power centers available in our local environment and we need only feel our way to or remember them.

Around the time of inferior conjunction, you can also choose to engage in the process of hierotopy or creation of sacred space. The king of spirit operates on the principle of Hamlet: "there is nothing good or bad but thinking makes it so." By far the most powerful way to create power and consecrate sacred space is through the expression of gratitude. By speaking our gratitude and making generous offerings we create sacred relations and sacred space. Thus, we must remember to use our hours and places of power not just to achieve things for ourselves, but also to give back. Giving thanks completes the circle and keeps power flowing through our lives.

Spiritual Practice: Illuminating the Sulphur
The Vision Quest Ceremony

You can learn to consult the soul of Hermes as an oracle in any of his sacred places such as crossroads, marketplaces, post office, libraries, or bus depot. Also, any place where symbols of Hermes are found can become sacred. These symbols include: the caduceus (the speaker or herald's staff), petasus (wide brimmed sun-hat), purses or pouches, musical instruments (especially stringed instruments and pipes), books, herma (a bust, sculpture, or mound of stones located near graves, boundaries, roads, or trails), wings, snakes (and other phallic symbols), roosters, sheep, tortoise, and ibis. Recognizing or bringing any of these items with you shows your knowledge and respect of Hermes, and serves to consecrate the space and aids in invocation.

Around the time of inferior conjunction, embark upon a vision quest to find and consecrate a personal power place sacred to Hermes. Remember that every event and person you meet on this journey is sacred, and must be treated with respect, humility, generosity, and thankfulness. Keep a record of your experience, because you can later mine this entire journey as if it were a dream, seeking out the deeper clues that reveal the correspondent archetypal contents. The heart of this quest is a consultation with the soul of Hermes himself.

To consult the soul of Hermes, first quiet your mind and enter into a state of reverence. Try to imagine the presence of Hermes nearby. When you feel any inkling of this presence, acknowledge the thrice-great god and give thanks for his many contributions to humanity. Then through prayer, humbly ask for his assistance in integrating your inner opposite. As you thank Hermes for his blessing, stop up your ears with your hands or earplugs. Slowly get up, leaving your offerings and as you exit the sacred place, trust that when you open your ears the first things you hear will contain an important message of your future. Be sure to mine this message repeatedly, until you retrieve all its many gifts.

Spiritual Practice: Illuminating the Mercury
Sacred Geometry, Mandala Creation & Meditation

Mandalas have a long history of as a tool for spiritual seekers. Creating a Mandala serves a kind of living prayer or artful meditation. The Mandala is said to represent the inner spiritual universe of the practitioner and is seen to serve not only a prayerful but also protective function, because the Mandala is a universal symbol of order, unity and wholeness. The sacred geometric order of the Mandala is a reflection not just of the organizing principles of the archetypal realms but can also be found in our solar system through the relationships of the planetary orbits to one another.[40]

Around the time of inferior conjunction, spend some time creating your own mandala. Start by simply drawing a circle and within the circle a triangle (similar to the earlier images of the triangle Mercury traces in the sky and of the tria prima with serpents). From there let your intuition take over and elaborate the mandala as you feel called, perhaps adding more interlocking triangles or pasting in images of your own personal spirit animals. Use whatever tools are available, adding color, using collage or found objects, or even using digital graphic software and images available online.

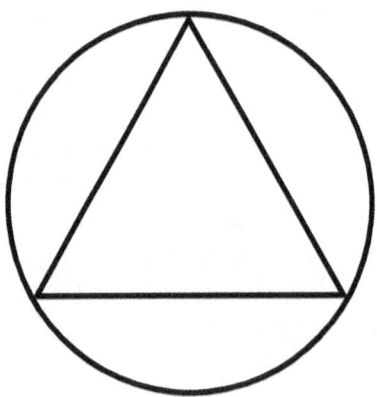

MORNING ELONGATION, RETURN, RUBEDO

The third and final time Mercury passes the thrice-crossed degrees, at morning elongation, Mercury is in the process of matching and surpassing the speed of the sun and thus crossing the threshold of return to the "normal" world where his average speed is faster than that of the sun through the sky. In alchemy, all transformations begin with fire. As Mercury became slower than the sun at evening elongation, the purgatory fires of the nigredo served to blast away the blackened impurities. During retrograde motion, as Mercury aligned with the sun, this purification allowed space for the new white light of albedo to be kindled in the soul. Now at morning elongation it is time for the rubedo, or reddening phase. Undergoing reddening in the fires stimulates the movement of fresh blood into the purified and ensouled matter, so that it may become whole and active once again.

The albedo phase represents the *heiros gamos* (sacred marriage) of the queen of soul, also known as the white queen, to the king of spirit, also known as the red king. However, it is in the rubedo stage that we pass from the initial wedding into a full blown working relationship between these two primary aspects. The rebis, philosophers child, or third consciousness which was conceived by the union of the white queen and red king during the albedo, must now be raised to maturity. In the laboratory this means the material must be re-heated. However, unlike the direct consuming flame of the nigredo, the flames of the rubedo are a gentler more controlled heat, similar to a bird sitting on eggs to hatch them. After the purging of the darker or unproductive passions in the nigredo, and the inner marriage of the albedo, it is time to re-engage the fiery passions toward more worldly goals once again. In real life, this simply means using the fires of inspiration kindled in the soul during albedo to produce the blood, sweat and tears necessary to make a dream real.

Jung says the white light of the albedo comes from the unconscious, as a dream does, and can be symbolized by moonlight heralding the coming of the rising sun. "The growing redness (rubedo) which now follows denotes an increase of warmth and light coming from the sun, consciousness."[41] Similarly, the third and final time Mercury passes the thrice-crossed degrees, he is in his morning star incarnation, shining high above the eastern horizon at sunrise. With the quality of light overcoming darkness, both sunrise and the east have always had obvious associations with new beginnings and venturing forth. Classically, Mercury as morning star was considered the guardian, "spearbearer," or scout of the King. Modern astrologers consider this a "Promethean" position. From his elevated position, Mercury has the ability to see clearly ahead toward far off goals and to make plans to reach them. Invigorated by his underworld visions of the future during the albedo, Mercury is eager to find their external correlates and make the "not yet me" or future self, into a corporeal reality.

If we are to experience the gifts of Mercury retrograde, every four months, as Mercury nears and passes morning elongation (labeled *return/rubedo* in the tables of the appendix), we must consecrate and become messengers of the future we have seen within us. After morning elongation, for the next two months or so, Mercury is again moving faster than the sun and so we can more easily move ahead unrestrained in the material world. In short, we must do the hard work and practice embodiment of that which is seeking to be born through us. There may be a temptation to heave a sigh of relief and then simply "return to normal" and go back to business as usual, living as we were before the interruption of the nigredo. However, this would deny the ultimate gifts of the descent into the spiritual and dangerously re-enforce the spirit-matter separation. The purpose and goal of the process is to bring something back from the spiritual realms and make it real. Thus, any attempts at returning to normal could necessitate a re-activation of the nigredo. However, rather than a return to normal, we should hope that whatever new parts of ourselves we have retrieved from the spirit world will result in at least a little chaos when introduced into our external world. It took a lot of internal chaos to achieve the light of the albedo, and so the appearance of some externally chaotic conditions at this point actually demonstrates that the inner chaos of the spiritual realms has been successfully transferred, that spirit and matter are in correspondence. If the heroine truly bears the gifts of wholeness, then by necessity the presence of this rare and precious quality should stir things up a bit. This may, and often does, happen completely organically, and when it does the wise alchemist will welcome and encourage the ferment.

The laboratory operation which corresponds to the beginning of the rubedo is fermentation. The harvested material (i.e., barley or grapes) is mashed to release its inner potentials (sugars), put into a solution and then a fermenting agent (yeast) is introduced. As the yeast consumes the sugars they release alcohol (spirit) and carbon dioxide, which results in a prodigious bubbling. Before this process was completely understood and mastered, the mashed grain solution was simply left out in the open air where it would eventually encounter enough airborne wild yeasts to get the process going. In lieu of awaiting a natural ferment, the alchemists would sometimes introduce manure. Similarly, in South America chicha beer is fermented by chewing or spitting into the grain. The resulting putrefaction temporarily re-activates the nigredo process, but this breakdown also eventually catalyzes the natural engines of fermentation and thus the rubedo. Psychologically, the lesson here seems to be that descending into the darkness of the irrational structures of consciousness and finding the light of soul therein is very hard work indeed, but ultimately only part of the battle. The presence of this new volatile spiritual energy (third consciousness) in our lives will naturally induce a ferment, which can appear at first as somewhat chaotic conditions. Ultimately, we must be willing to sacrifice the new spiritual gift we have found to the material world, allowing it to be broken down and fermented, in order to release its purest essence.

Jung explained the presence of this kind of positive or productive chaos by way of a process called compensation. Jung believed that, like the body, the psyche must also be self-regulating toward equilibrium, and indeed ultimately toward wholeness. Jung called the source of this urge toward psychic order and unity, *the Self*. The archetype of the Self is transcendent and lies outside the individual psyche, in the collective unconscious, and yet its manifestations appear individually.[42] We can think of the archetype of the Self as a sort of demiurge, a universal artisan-like creative force similar to what masons, gnostics and indeed Hermeticists call the "great architect of the universe."

For Jung the process of human development is life-long. The archetype of the Self is constantly introducing chaotic elements of the unconscious into our lives in order to compensate for or give balance to our conscious preferences, which tend to become fixated and one-sided. The divine volatility of the unconscious is a necessary compensation to the natural fixity of the conscious ego. In order to keep us growing, the Self may introduce this unconscious material to us in a series of very symbolic dreams or imaginative fantasies. We may also find that this process of making the unconscious material become conscious happens by way of synchronicity, that is, by meaningful coincidences occurring in our conscious life that seem to mirror or be in correspondence with the whisperings of our unconscious dreams and inner imaginings. As above, so below *and* as without, so within; the conscious/unconscious and inner/outer lives often mirror one another and the wise alchemist must learn to become aware of and follow along with this process.

If we are conscious of and responsive to the whisperings of the great architect in the form of dreams and synchronicity, then the chaotic contents of the unconscious can often be assimilated in a relatively gentle or mostly harmless fashion. Many who have be-

VITRIOL, from Basil Valentine's *Chymical Wedding*, mid seventeenth century. *Visita interiora terræ; rectificando invenies occultum lapidem* ("Visit the interior parts of the earth; by rectification you will find the hidden stone.").

come aware of the calling of the Self seem to take great pleasure in the seeming game of psychic cat and mouse that enables one to stay on course toward the ultimate wholeness, unity, and divine order which the great architect desires for us. However, if we are unconscious of or unresponsive to the more subtle manifestations of the unconscious in our lives, then it may need to take on more and more gross forms, in order to achieve the necessary compensation. In fact, even seeming disasters can be attributed to the compensatory nature of the Self, with the long-term result being a positive outcome that could not have been found by the conscious ego without intercession.[43]

Jung used the term *enantiodromia*—meaning literally "running counter to"—to explain the more gross manifestations of the Self and the seeming intrusions of the unconscious into our lives. Just as Mercury runs counter to its normal direction during the retrograde, it is only natural that we encounter unconscious material that runs counter to our conscious preferences during this period. The less able we are to recognize and accept the more subtle manifestations of the Self's compensatory forces, the more psychic pressure is built up. Over time this pressure can lead to dramatic consequences.

Using a metaphor from lab alchemy, without an appropriate venting for the gasses of fermentation, a closed container will eventually explode. On the other hand, if the fermentation is happening in an open system (i.e., fermentation with wild yeasts in vats exposed to the open air) and the results are not captured in a timely fashion, then the fully fermented beer or wine will be subsequently subjected to bacteria and spoiled. We must learn to successfully contain, concentrate or focus, and eventually harvest in a timely fashion the chaotic energies of the unconscious in our lives, in order to move to the next step of the rubedo phase.

"Separate the subtle from the gross, gently and with great ingenuity," advises the Emerald Tablet regarding the process of distillation. Just as the lab operation of distillation requires a continuous cycle of gentle heating, condensation, and then re-heating, we must also learn to enter into a cyclic and rhythmic relationship with the unconscious in order to obtain the presence of a more subtle and concentrated spirit in our lives. After the rigorous ferment ensuing from the introduction of the new third consciousness first glimpsed in the albedo, we must continue to refine and perfect our new vision until its presence becomes more and more subtle and penetrating. Following the model of the Emerald Tablet itself, eventually we need to find a way to encapsulate the entire process of our journey from nigredo to albedo and through rubedo into a single image or brief short story. This encapsulated essence of the art of transformation can be seen as an individualized representation of the ubiquitous and yet elusive *philosopher's stone*, the elixir by which the heroine passes on the gift of wholeness.

Spiritual Practice: Uniting the Salt, Sulphur & Mercury
Creation of the Personal Philosopher's Stone

After a rite of passage, we must take care to sufficiently demarcate for ourselves a boundary between the former self and the newly transformed identity. By encapsulating the spirit of our new identity within an image, song, poem, or the like—we are creating not just a memento or keepsake to serve as physical reminder of our newly emerging future self, but also a magical talisman which will serve to draw us toward our future and keep us from falling back into the old habits originally discarded in the nigredo.

Around the time of morning elongation, spend some time creating your own talisman. This could take the form of a new medicine pouch, which carries various herbs, crystals and representations of power animals, which symbolize the new medicine you wish to carry and embody in this world. You could also take your mandala practice to the next level, creating a personal image of the Self and accompanying it with a song, poem or short passage which consecrates and encapsulates the emergent vision of your future self.

Whatever form you use to create your personal philosopher's stone, remember to give thanks and all honor for this gift to the thrice great Hermes—and remember also that you are now a messenger of the messenger God. As you gradually become able to embody and reveal your personal hierophany to the world, you are merely recapitulating the original hierophany of Hermes. Therefore you must take care and be on guard not to let selfishness and ego attachments take over your relationship to your sacred stone, for it is a gift meant for the collective, and you are but the vessel through which it flows.

A BRIEF EXAMPLE OF THE TRANSFORMATIVE PROCESS

While I have given a separate description of the temporal quality of the three related events (evening elongation/nigredo, conjunction/albedo, and morning elongation/rubedo), it is important to remember that they are in fact a "differentiated unity." By definition, each separate event is inextricably connected and related to two others. That is to say, in real life each of these three events can and often do contain shades, nuances, and influences of the other two. Furthermore, it is important to remember that spiritual processes do not always strictly conform to linear time. In other words, even while we may feel ourselves to be in the rubedo stage of transformation—that is, moving forward with the spiritual vision gained at conjunction/albedo—some elements of the nigredo stage may still be present. Thus, further releasing of the "no longer me" shadow may still sometimes be necessary during the rubedo, before the "not yet me" of the rubedo can be fully claimed.

A classic example of this process is the release date of second best-selling album of the twentieth century. The Eagles released their *Greatest Hits* (1971–1975) on February 17, 1976 with Mercury at 2° Aquarius and near maximum elongation as morning star. Aquarius belongs to the naturally communicative air element, and it is also where we find the very bright star Altair of the constellation Aquila, the Eagle. So this particular zodiacal degree is a quite fitting and powerful one to be activated for this particular band to release an album! The intrinsic nature of morning elongation is also felicitous for this particular album because at morning elongation Mercury is not only in the same degrees as in the previous conjunction, but also in the same degrees as the previous evening elongation. Therefore, this event is effectively a summation of all three occurrences.

As a greatest hits album it is by definition a statement of that which is "no longer me" and has an obvious connection with the past (evening elongation/nigredo). At the same time, greatest hits are also by definition the songs that represent the closest alignment of the band with public tastes, representing the clear spiritual insight possible via conjunction/albedo. Finally, since it represents a specific period of development for the band, which has now been fulfilled, this album also activates the freedom of morning elongation/rubedo and unlocks the future. Because the band's past has been formally encapsulated, the band can now pursue that which is "not yet me" without being chained to a previous identity.

Indeed, while the band was releasing this compilation, it was also undergoing very substantive personnel changes. The four founding members of the Eagles signed with David Geffen in September of 1971, during the last part of a Mercury elemental year of earth (and as the Mercury elemental year was transitioning from earth to fire). By the time their first greatest hits album was released, the Mercury elemental year had moved through fire, water, and air, and was about to return to earth. This six- to seven-year return to the Mercury elemental year of birth represents what I call the "long form" of Mercury's transformative dance, and usually indicates some kind of completion. Like all completions, it can signify a return to ones roots on the one hand, or the beginning of a brand new era on the other. In the case of the Eagles, it was the latter. As their sound was moving from a mostly country (earth) influence to more rock and roll (fire), founding member Bernie Leadon left the group (December 1975). Leadon's solid multi-instrumental talent (earth) was replaced by the volatility of Joe Walsh (fire), and the first album featuring Walsh in the line-up, *Hotel California,* became not only the band's best-selling studio album, but also a rock and roll classic. Through capturing the transformative power of Mercury's morning elongation, the Eagles were able to transform themselves from country-rock pioneers into pure rock and roll legends.

Chapter V
Elemental Magic

The Sun is its Father (fire);
its Mother the Moon (water).
The Wind carries it in its belly (air);
its Nurse is the Earth (earth).

—HERMES TRISMEGISTUS
IN THE *EMERALD TABLET*[44]

THE FOUR ELEMENTS of fire, water, air and earth actually predate Greek horoscopic astrology by at least a few hundred years. In these early discussions on the nature of the universe, amongst the pre-Socratic philosophers there were those called *monists*, who had assorted ideas about which one of the elements was primary. Various theories were proposed by which one primary element is supposed to have generated the others. Around 500 BCE Empedocles of Acragas offered a novel solution to these arguments, saying that the universe consists of a mixture of what he called the four "roots."[45] In Empedoclean cosmology, the four elements together are primary, and all material reality consists of these four roots mixed and separated in various proportions, like the colors on a painter's palette. The mixing and separating is achieved by the eternal cosmic forces of love and strife.

Empedocles' model is remarkably both proto-astrological and proto-alchemical. We find here not just the foundation of the four elements theory, but in the waxing and waning of the forces of love and strife, we can also see reflected the three primary cycles of astrology—daily, monthly, and yearly—which all oscillate between light and dark. Furthermore, Empedocles not only identifies various Greek deities with the four elements (Zeus, Hera, Hades, and Persephone in fragment 6) but goes so far as to also equate the four elements with natural phenomena (sun, earth, sky, and sea, in fragment 22). Thus we can see here the classic above/below correspondence, which later becomes ubiquitous in the alchemical literature.

The four elements have themselves now become ubiquitous, forming the bedrock of many symbolic systems and mythologies. As such they have had a profound impact on western esotericism (especially astrology, alchemy, magic, and tarot). Further developed by such notables as Plato, Hippocrates, Galen, Aristotle, and Jung, the four elements are also present in the more "scientific" thinking via the four humours and temperaments of early Hippocratic medical theory, as well as in modern Jungian psychological typology. The four elements have thus provided an archetypal foundation for a wide range of disciplines throughout history.

Perhaps this ubiquity is due to the fact that the four elements are inherently empirical or easily observed. For instance, Empedocles showed that the existence of even the relatively abstract and hard to see air element can easily be demonstrated by simply turning a bucket upside down, submerging it in water and noting the buoyancy caused by the air trapped inside and the bubbles escaping as the bucket is turned back upright. Thus, even unto this day, the four elements still prove remarkably practical as categories for classifying matter in various states of density—i.e., solid (earth), liquid (water), gas (air), and plasma (fire). Similarly, we can use the example of burning wood again to see the immediacy of Empedocles' assertion that all matter is made from a mixture of all four elements. If we put a piece of living green wood on a hot fire it will first begin to weep liquid from the cut ends; therefore wood contains water. With continued heat the water will turn to steam and vapors; therefore wood contains air. After becoming dry, the wood burns; therefore wood contains fire. Finally we are left with ash, demonstrating that wood contains earth.

We can classify the elements in various ways. Air and fire are relatively less dense and more volatile. Because they tend to rise, moving up and out, fire and air are connected with the king of spirit. When we rise with spirit, we get the big picture and overview of life, or find moments of personal freedom or accomplishment, the ultimate of which Maslow called "peak experiences." Water and Earth are relatively more dense and fixed. Because they tend to move down and in, water and earth are connected with the queen of soul. All the riches of water and earth are found beneath the surface. The soul wants us to grow down and become deep like a river canyon, revealing the complex layers of our inner nature, or rooted like a tree, and thus able to withstand the storms of the more volatile elements.

ELEMENT ORGANIZATION	FIXED Rational Impersonal Objective Word	VOLATILE Irrational Personal Subjective Image
VOLATILE Spiritual Incorporeal/Energy Assertive/Electric Yang/Outward	Air △	Fire △
FIXED Yin/Inward Receptive/Magnetic Corporeal/Matter Instinctual	Earth ▽	Water ▽

Somewhat ironically, it was Aristotle who first introduced a transformational dimension to the four elements. Aristotle took the four qualities described by earlier philosophers such as Anaximander, (hot, cold, moist and dry) and demonstrated through them the capacity of Empedocles' four elements to cyclically change one into another. For Aristotle, each element possessed two of these primary qualities, but one predominates. In other words, each element has a dominant quality and secondary quality—i.e., fire is primarily hot and secondarily dry, or hot *becoming* dry. When arranged in a square, we can see that elements adjacent to one another, share a quality, and so the idea of becoming introduces a circular or cyclical motion. Using the analogue of the four seasons, we can see that spring is wet becoming hot (air), summer is hot becoming dry (fire), autumn is dry becoming cold (earth) and winter is cold becoming moist (water), as in the figure on the opposite page.

This idea of a cyclic becoming or generation of the elements became a serious fascination for the alchemists, who ultimately sought to achieve the *union of opposites* (for example, fire and water), though for Aristotle, this would have seemingly been a philosophical impossibility. To understand how Aristotle's idea of the elements becoming one another was transformed into the seeking of a unity of opposites by the alchemists, we need to consult what is known as "the square of opposites," a fundamental teaching tool for renaissance era magic. Via the square of opposites, we can see that adjacent elements share a quality, and this sharing is the key to the process of becoming or transformation. And, perhaps to Aristotle's dismay, we can easily see the "magic" of this process in (what have become) simple everyday events. For instance, both water and air are moist; so that when we replace the cold quality of water with heat, it transforms into its gaseous state by boiling. Likewise, if we replace the cold quality in earth with hot, it will eventually combust and produce fire, just as heating a piece of dry tinder enough via friction results in flame.

Another way of seeing the transformation of adjacent elements on the square of opposites might be that fire devours earth (via combustion) and air devours water (via evaporation). Going the other direction, we can also see that air feeds (or is moved by) fire and that water feeds (or moves through) earth (osmosis). Whichever direction we take, the transformation of elements is (relatively) easier when they are adjacent on the square of opposites. This process of transformation by moving from one adjacent element to another around the outside of the square of opposites was called "making the quadrangle round" by the alchemists.[46]

Recall again that each year Mercury's triple alignments all activate three signs of the same elemental triplicity. The element in which Mercury's retrogrades were happening during the year of your birth thus defines your personal elemental preference. Think of how a wine is defined by its vintage: all wines bottled by a vintner in a particular calendar year share the same general characteristics. Similarly, the Mercury elemental year of your birth defines the vintage of your image making, information processing, and spiritual transformation faculties. Though perhaps at first somewhat unconscious, you are here to make a contribution to the collective by forming new answers to a fundamental human theme, need or question as symbolized by this element, and by learning to transform into its opposite element.

The sky keeps turning, and Mercury's elemental theme changes every 16–24 months (after four to six retrograde periods), moving

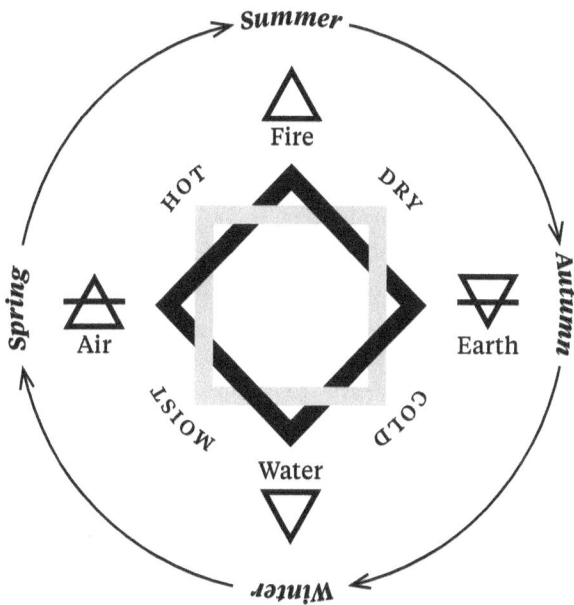

completely through all four elements from fire to water to air to earth over a period of six to seven years (after 19–22 retrogrades).[47] This is also the sequence of elements mentioned in this chapter's opening epigraph, quoted from the Emerald Tablet. The purpose of this passage is widely seen by alchemists to refer to a series of operations by which the initial transformation of a substance is achieved. The sequence also happens to precisely correspond with the passage of Mercury's retrogrades through the four elements over the course of six to seven years. As above, so below.

When viewed through the square of opposites, what this cycle means for the Mercury elemental years is that the transitions from water to air years, and from earth to fire years are the most natural or immediate, while the transformations from fire to water years and air to earth years are more difficult, perhaps requiring intermediate steps. Because they do not share a quality, transformation between elements opposite to each other on the square of opposites is much harder. Fire cannot be directly transformed into water, it must first be transformed into air or earth. Curiously, this is very much like the doctrine of shadow and anima in Jungian depth psychology. One cannot simply proceed directly to the masterpiece of integrating the inner opposite, but must first go through the apprentice stage of integrating the shadow. Therefore, the Mercury elemental year transitions from fire to water years, and from air to earth years are the most difficult, because they require an intermediary step, or two transformations. It could be inferred from this that the more volatile elements, air and fire, also contain a more volatile shadow and that their shadow is of a nature which somehow requires more immediate work.

Both Empedocles and Jung appear to agree that humans are more whole when they are able to understand and express themselves through a relative balance of all four elements. Nevertheless, it is also true that we are all fated to be born during times in which Mercury is demonstrating a decided preference for spending much more time in only one of the four elements. Likewise, Jungian typology posits that we each naturally have an unconscious preference for one of four modes of being (thinking, feeling, sensation, intuition). Thus, in order to become whole, we must undergo the journey of transformation and conscious integration of our internal opposite. In fact, in order to experience and integrate all four elements, it appears we need to take this road less traveled at least twice. That is, we must travel both paths and both directions on the square of opposites between our element of birth and its opposite.

Fortunately, Mercury gives us this opportunity by moving through the four elements every six to seven years. We find that the same universality of a threefold transformational process is still present in the fourfold process. Whether you want to apply through alchemy (production/fire, adaptation/water, expansion/air, retraction/earth), take a Jungian typology test (intuition/fire, feeling/water, thinking/air, sensation/earth), use the native Medicine Wheel (independence/fire, belonging/water, generosity/air, mastery/earth), or simply note the levels of expression of energies in your life (spiritual/fire, emotional/water, mental/air, physical/earth), it is likely that you will find an abundance of one element, and therefore the lack of another. Again, we can see the six- to seven-year cycle of the Mercury elemental year as an opportunity for finding a more balanced expression of the four elements in our lives.

MAKING THE QUADRANGLE ROUND

The Mercury elemental year changes every 16–24 months (after four to six retrograde periods) and moves completely through all four elements from fire to water to air to earth over a period of six to seven years (after 19–22 retrogrades). In order to keep growing and evolving, we need to experience the re-alignment of our "three selves" at least once during each of the four Elemental Years. Using this long form of the process of transformation, and "making the quadrangle round" by deeply and consciously experiencing the cycle of four elements, prepares us for initiation into the most profound transformation into our inner opposite, which is a rare and special occurrence that may take many attempts before reaching fruition or completion.

The return of the Mercury elemental year to your birth marks both the ending and beginning of the cycle, the place where the snake swallows its own tail forming the ouroboros. This is a time when you can be more yourself and when your natural inborn predilections for image making and communication are generally more active and welcome in the social and collective spheres. However, it is not a time to simply coast or take it easy. These returns mark a significant stage of life development that only occur every six to seven years, and thus present a key opportunity. This can be a time of completion and important new beginnings, or it can represent a return to your roots and another iteration of a lifelong theme after a period of explorations. Each of these returns has its own unique developmental tasks, where the basic theme, need or question of your birth element are viewed and answered with new perspective, and these are outlined in the following chapter. In between the returns to the birth element, we must practice becoming whole by allowing ourselves to be transformed by each of the elements in its turn.

Fire signs (Aries, Leo, Sagittarius)

Fire is hot becoming dry and is the most volatile and least dense of the four elements. The basic nature of fire is to rise (hot) and separate (dry). Thus fire serves to answer the basic human needs for individuality, energy or spiritedness, intuition or instinctive knowing, faith and optimism. Fire changes and moves quickly, constantly seeking out and devouring new material and has the ability to destroy or leap across barriers. Likewise during fire years we need to find new sources of creative fuel and inspiration and learn to use this fire to burn off any accumulated detritus in our lives and to purge falsity in ourselves. Fire years are great for realizing our uniqueness and individuality, creatively or spontaneously going after what we want, learning to become competitive and seeking to reach the top, seeing and creating our future and illuminating our highest spiritual truths.

For those born in water years, fire is your natural inner opposite and so this transition is the most difficult. You may need to practice earth or air modes of being first, but ultimately becoming fiery completes you and brings you and your universal message to new shores. For those born in air years, fire shares the quality of heat but you must replace moist with dry to become fiery. This means separating yourself and breaking attachments in order to assess which connections really serve your spirit. For those born in earth years, fire shares a quality of dryness but you must replace cold with hot to become fiery. This means becoming more active and spontaneous, choosing action before evaluation, and sacrificing capital to achieve things.

Water signs (Cancer, Scorpio, Pisces)

Water is cool becoming moist and is both fixed and fluid. The basic nature of water is to sink or go within (cool) and to unify (moist). Thus water serves to answer the basic human needs for emotional depth and breadth, developing sympathy, empathy and compassion, attunement to and faith in the unconscious and its messages, and to assess quality through feeling values. Water changes temperature very slowly and yet can change position by flowing very rapidly. Water has no shape of its own but takes on the shape of its container. Likewise, during water years we need to learn to become malleable and flow with life, absorbing the universal messages around us, and allowing them to dissolve any rigidity in our lives. Water years are great for quiet inner and preparatory work, creating new channels for emotional expression or attunement with the unconscious, and for adjusting our feeling values and assessing our state of emotional happiness and contentment with our lives.

For those born in air years, water shares the quality of moistness but you will need to become cool to become watery. This means slowing down and practicing the Taoist state of Wu wei which means: "action in non-action." By practicing non-action and non-attachment to outcomes, you can allow all things to settle and coalesce of their own nature, and you will more easily see the natural harmonious connections and discard those that are false or filled with strife. For those born in earth years, water shares the quality of coolness but you will need to become moist to become watery. This means putting aside your natural skepticism and detachment, and relaxing your critical faculties. Allow yourself to feel connections between things, especially when they do not make rational sense. For those born in fire years, water is your natural inner opposite, and so this transition is difficult. You may need to practice earth or air modes of being first, but ultimately fire forms new lands in the ocean of water.

Air signs (Gemini, Libra, Aquarius)

Air is moist becoming hot and is the most naturally communicative of the elements. The basic nature for air is to rise (hot) and to unify (moist). Therefore air serves to answer the basic human needs for intellectual activity and theoretical unity, cheerfulness and generosity, belongingness and cooperation with others, and to quickly understand and adjust to diverse ideas and people. Air flows like water, yet can change temperature relatively quickly. The ability to listen and gauge interest is just as important as the ability to broadcast and disseminate. Likewise during air years we can learn to flow with social trends and change temperature to either match or initiate changes in our environment. Air years are great for marketing, advertising, and initiating information projects like websites. Making or tending social connections with like-minded individuals is also important during air years. Remaining flexible, changeable and adaptable and quickly moving to match changes or respond to synchronicity is important in air years.

For those born during earth years, air is your natural inner opposite, and so this transition is difficult. You may need to practice fire or water modes of being first, but ultimately the dream of flight is your deepest yearning. For those born during fire years, air shares the quality of heat but you will have to learn to become moist to become airy. This means curbing your tendency to go it alone, and joining with like-minded allies before adventuring. Testing connections first, to see which ones are most important or enduring, can also be valuable. For those born in water years, air shares the quality of moistness but you will need to become hot to become airy. This means becoming more active, embracing and flowing with any sacred agitations, rather than seeking placidity.

Earth signs (Taurus, Virgo, Capricorn)

Earth is dry becoming cool and is the most fixed of all the elements. The basic nature of earth is to sink or go within (cool) and to separate (dry). Therefore earth serves to answer the basic human needs for tactile understanding, structure and discipline, constancy and firmness as well as natural diversity of interest and practical skill building. Earth provides the boundaries to both sea and sky and also naturally reveals structures, strata and classifications. Likewise, during earth years we can learn to understand the boundaries, rules, and structures so well that we can eventually bend them to our advantage. Earth is made of diverse minerals, and so earth years are great for skill building, collecting and blending the best practices of those we respect. Paying attention to accounting, managing resources, and developing efficiency are important earth themes as well.

For those born during fire years, earth shares the quality of dryness, but you will have to become cooler in order to become earthy. This means paying more attention to the rests between movements and the silences between sounds. It is only through their combination that music is created instead of mere noise. For those born during water years, earth shares the quality of coolness, but you will have to become drier to become earthy. This means cultivating a healthy respect for boundaries and structures and living within or flowing through them instead of under, over, or around. For those born during air years, earth is your natural inner opposite, and so this transition is difficult. You may need to practice fire or water modes of being first, but ultimately your deepest yearning is the dream of a solid, stable, and practical home that you can return to after your wanderings.

Both the return to the Mercury elemental year of birth, and the opposite element to the birth year on the square of opposites, contain immense potentials. At the return of the birth year, one has the ability to use their natural inborn strengths and preferences to advantage, perhaps finding new ways to exploit them and also occasionally repeating previous successes, though perhaps taking it to another level. During the intermediate years, when Mercury is making the triple alignments in elements adjacent to the birth year element, one has the chance to make the preliminary transformations necessary before one can achieve the hardest task—the integration of their opposite element. Until one has learned the lessons of integrating the two adjacent elements, Mercury elemental years of the opposite element could be experienced as especially challenging. Nevertheless, it is ultimately through integrating one's inner opposite that the greatest growth and wholeness is experienced.

The process of "making the quadrangle round" through the experience of the Mercury elemental year's movement through the four elements can be illustrated in the story of how this book came to be. I am born in a Mercury elemental year of air, so I have always had a natural talent for intellectual activities like writing and public speaking, though I was never entirely comfortable expressing these talents through the socially acceptable channels, such as academia. I felt rather constrained by these options and sought alternatives. Additionally, the opposite element to air is earth. Ultimately, I needed to experience more practical forms of writing, the every-day kind that most people actually read. Beginning in 2003 (an earth year) I began to have my astrology articles accepted for publication in various journals. I also began my career as a lecturer and public speaker on astrology. Over the next seven years I gained experience researching, writing, publishing, and speaking in local, national, and international contexts. After seven years of experience and a round through all four elements, my "lesser stone" was complete, and I had become a respected professional in my new community.

During the next earth years, I received another growth challenge. Until then I had mainly spoken in my home state of Virgina and surrounds. In 2010 I received invitations to speak at major conferences in Boston and New York. At the time, my favorite subject was Venus, but someone was already scheduled for that topic, so I agreed to give a talk on the cycles of Mercury instead. At this stage, my knowledge of Mercury was barely at the point where I thought I had something important to say, and so I felt forced to deepen my studies. I completed my first lecture tour, through the mid and southwest that year, sharing this newfound knowledge on the cycles of Mercury.

During the subsequent fire years, I soon became consumed by all things Hermetic, including further investigations into alchemy (which is all about transformation via fire). One day, after I had been out photographing the planet Mercury, I was back at my desk meditating upon a copy of the Emerald Tablet that I always kept there. Suddenly, I received an insight that the pattern in the text (i.e., Above, Below, Above) actually matched my astronomical observations. I called my alchemy tutor to reveal my discovery of the microcosmic triple alignments (the forty-day "short form" of the transformational cycle), and in the midst of explaining these, I also realized that the long form (six to seven year cycle of Mercury retrograde through the four elements) matched another later passage in the Emerald Tablet (quoted at the start of this chapter). The sudden transformation of my understanding via these intuitive glimpses of truth as well as the immense excitement and ultimate growth that they generated are the perfect correlate for the experience of the fire element. I had broken apart the secret code (*solve*).

During an introspective winter in the water years of 2012–13, I took up the arduous task of putting together (*coagula*) the first version of the tables in the appendix. Oddly, after completing them I put them away and later almost forgot about them. I connected with my publisher during the following winter, and while discussing an unrelated publishing project, the subject of these tables came up. During the ensuing air year of 2014 (the seventh return of my birth year element—see chapter six) we both agreed we wanted them published.

It was during the process of writing the book proposal and explaining the premise, that I realized there were still deeper truths I needed to convey, and that this book is indeed only the first installment of a trilogy. So, the idea of *Hermetica Triptycha* was born during my Mercury elemental year return. During the ensuing earth period, I was again challenged to bring the theory (air) down to earth, and spent many hours researching the examples, writing and re-writing versions of this book. Now in 2017, after "making the quadrangle round" by studying, writing, and speaking specifically on Mercury through a complete cycle of all four elements, the first volume of my *ars magna* (great work) is complete.

PRACTICAL THEURGY
Magical Aid for Transformation via the Four Elements

By engaging in magical or ritual work we can facilitate both the short (forty-day) and long (six- to seven-year) forms of the transformational process. I have found that the simple act of keeping and tending an altar with sacred items representing each of the four elements can be a very rewarding spiritual practice. For instance, one year for Christmas my brother-in-law gave me some Canadian Silver Maple Leaf coins. I wanted to honor this gift, so I put them on the earth section of my altar. Soon I began to receive interest from students and clients in Canada, and in synchronicity my phone company decided to expand service there around the same time, so I had a nice new income stream. Sometimes transformation into the opposite is easier when we recognize and allow its natural catalysts to act as go between. In order to assist in "making the quadrangle round," we should learn to pay special attention to the objects and processes of each element in its turn, especially our birth element and its opposite. As the Mercury retrogrades activate each element, watch for, welcome and flow with its various manifestations and appearances in your life during its respective years.

Fire signs (Aries, Leo, Sagittarius)

Fire objects include things like: candles, incense burners, and wands. Fire magic can be as simple as lighting a candle, burning incense, or combusting ritual objects that symbolize "no longer me" in a safe place, like a fireplace. More complex fire operations include calcining herbal material, reducing it to ash containing basic mineral salts, or fermenting barley or grapes. Psychologically, fire represents the creative imagination and our sources of inspiration. Archetypally, fire is our warrior persona, the part of us that rises above our fears and fights for what we need and want, striving to reach the top.

Water signs (Cancer, Scorpio, Pisces)

Water objects include things like: cups, chalices, and oils/potions/elixirs. Water magic can be as simple as spending time by a stream, lake, or beach or using liquid flower essences. More complex water operations include dissolving herbal material in alcohol, or distilling water or spirits to purify them and extract the essential oil of the plant. Psychologically, water represents the unconscious and working with one's dreams, feelings, and values. Archetypally, water is our altruist persona, the caretaker part of us that nurtures and heals.

Air signs (Gemini, Libra, Aquarius)

Air objects include things like: journals, feathers, and cutting blades. Air magic can be as simple as prayer, affirmations, or journaling. More complex air operations include any kind of separating or sifting of entangled or enmeshed materials, whether physically with aid of a cutting or sifting instrument or "sorting things out" through processes like spiritual contemplation, meditation, therapy, or academic study. Psychologically, air represents our spiritual mind as well as our social connections. Archetypally, air is our wanderer persona, the questing part of our nature.

Earth signs (Taurus, Virgo, Capricorn)

Earth objects include things like: coins, crystals, and jewelry. Earth magic can be as simple as ritual adornment, ritually prepared meals, or the use of homeopathic and herbal remedies. More complex Earth operations include any mixing and combining of things to create something new. For instance a spagyric is an alchemical essence combining the *tria prima* of Salt (mineral salt of an herb), Sulphur (essential oil of an herb) and Mercury (spirit or alcoholic tincture of an herb). Psychologically, earth is about sensation or tactile understanding and practical skill building. Archetypally, earth is our magician persona that collects and blends diverse ingredients, synthesizing them into something new.

ADVANCED THEURGY: Magical Talismans via the Stars

Though it is beyond the scope of this book to explain or instruct in proper theurgy, I have nonetheless included in the ephemerides tools for the advanced practitioner. Along with tracking the triple alignments of Mercury through the four elements, I have noted the activation of the other "wandering stars" or planets, and the fixed stars activated and their constellations have also been noted, as these were traditionally used for the creation of magical talismans (see, for instance, Agrippa's *Three Books of Occult Philosophy* or the volume known as *Picatrix*).

Chapter VI
The Twelve Elemental Returns

> *"Make of a man and woman a circle; then a quadrangle; out of this a triangle; make again a circle, and you will have the Stone of the Wise."*
> —MICHAEL MAIER, *ATALANTA FUGIENS*[48]

SO FAR WE have seen how the Mercury elemental year illuminates the deeper field and context within which any birth or magical moment occurs, by way of the houses or places where the pre- and post-natal retrogrades occur. We have also discovered how both the triple alignments of an individual retrograde (the forty-day "short form" cycle) and the six- to seven-year cycle of through the four elements (long form) offer us extremely useful maps and instruction for navigating and completing the transformational journey. Even if we were to stop here and never explore or practice further, I am certain this book will still prove a most valuable aid in the hands of sincere seekers, alchemists, magicians, and especially practicing astrologers.

It is also my most fervent hope that this volume may ultimately inspire and enable an even broader and deeper lifelong study of the transpersonal and collective dimensions of planet and archetype of Mercury. If undertaken by enough individuals this may even facilitate a reintegration of the Hermetic principles and worldview within the practices of the as now mostly separate and splintered fields of astrology, alchemy, magic, and western esotericism. For far too long, the planet and archetype which inspired the awe and reverence of alchemists and mystics via the *Hermetica* for thousands of years has often been routinely dismissed in discussions of the collective or transpersonal in modern astrology. I see this is as a tragic situation that must be remedied, for once we look beyond the forty-day triple alignments, the elemental triplicity activated each year, and the six- to seven-year cycle through all four elements outlined thus far, we see another important pattern. It is most important to stress and understand that not only do the triple alignments occur in the exact same *degree* for each retrograde, but that Mercury will return to *that very degree* once every 79 years.

What can be the implications of this important recurrence? Since Mercury has such a precise, long-term cycle, does this not make it evident that Mercury also has a collective or transpersonal dimension, akin to the slower, modern planets beyond Saturn? I have already demonstrated that Mercury has *several* collective or transpersonal cycles and dimensions, and yet this 79-year cycle seems to dwarf all the others. This discovery puts the notion of the transpersonal realm being accessible mainly or exclusively through the modern outer planets into question. After all, did not Galileo's discoveries about Jupiter's moons and the phases of Venus precede the discovery of the modern planets? Since it is only through first accessing transcendent wisdom that we ever came to discover the outer planets, they logically cannot have exclusive dominion over the transpersonal!

I believe it is time for us to learn to look at this scale of planetary motion. I aim for these larger patterns in Mercury's cycle to become common knowledge, and for this truth to become evident, widely accepted and taught within the greater astrological community: that because, like Mercury, *all* the so-called personal planets, including the sun and the moon, also have long-term cycles, then they too must also have collective and transpersonal dimensions. From our study of alchemy we know that it is only through the painful and painstaking work of removing the outer corrupted layers of falsity that the pure original substances can be arrived at and extracted, and then recombined to recreate the original "quintessence" from which all matter ultimately derives. This is true at all levels—for individual people, their works, and their communities.

Thus, as we have done here with Mercury, it is only through carefully attending to and understanding the other so-called personal planets as visual phenomena, and exploring the other complex intricacies of their astronomical movements and cycles, that these collective and transpersonal dimensions and their corresponding meanings can be revealed. Therefore, as a collective contribution to the field of astrology, the most important thing this volume demonstrates is the existence of an entirely new frontier that must be explored, expanded, and expounded upon so that our community can be healed, whole, and deserving of reclaiming its previously exalted place in society. Of course, this vision fits with my natal Mercurial trigon being made up of the heroic houses of the Identity Trigon, houses one (pre-natal), five (post-natal) and finally nine—the house of far flung horizons and theoretical unity.

I mention this need to transcend and heal the imperfections of the community here mostly because it is precisely through this long term level of seeing and thinking that we can access the most transcendent of Mercury's cycles—for after all, seventy nine years is a lifetime. The ultimate goal of the heroine's journey is not simply to find the grail or elixir for her personal transformation, but rather also to bring it back and use it to heal the sovereign, community, and kingdom. The only way this can be done is through surrendering the work back to the work, by continuously re-entering the transformative journey of "making the quadrangle round," and making it into a lifetime commitment. In this way the hero or alchemist can achieve ever widening circles of perfection, with the liquor of their souls being transformed from simple beer or wine into a more sublime brandy or whiskey and ultimately the clear pure quintessence of wisdom. I believe this is what fifteenth-century alchemist George Ripley was intimating through the complex and convoluted layers radiating out from his "Ripley's Wheel"—an advanced version of the "square of opposites," which we used in the previous chapter.

Ripley's wheel was loaded with correspondences. Everything from the wheel of seasons, the four cardinal directions, the four elements, four qualities, four bodily humours, and signs of the zodiac (divided into the four elemental triplicities), with each quarter accompanied by verses and scriptural references. These were arranged such that they radiated outward in concentric circles, much like astronomical drawings of the heavenly spheres at the time. Since alchemy was often referred to as a kind of inferior or "lower astronomy," the concentric circles of Ripley's wheel can, and indeed should, be seen as a terrestrial analogue for the planetary spheres (as above, so below).49

Furthermore, because Ripley also includes four grades of perfection (origin, imperfection, perfection, and *plusquam perfectum* "more than, or super perfect"), his wheel can be seen as a context for explaining how the natural perfection of the celestial spheres can be generated by the alchemist within the sublunary, inferior, or lower sphere of the earth and four elements. Thus, we can see that Ripley's *Coelum philosophorum* quite literally depicts an alchemical cosmos: a true "lower astronomy" that describes the generation of heavenly perfection from various mutable, terrestrial elements. Ripley's wheel illustrates how a square of paired, contrary qualities may, *by continual rotation through the various levels,* eventually generate a fifth, "more than perfect" substance: the original quintessence also known as the "philosophers stone."50

SQUARING THE CIRCLE

Once this understanding of the process of perfection via Ripley's Wheel has been achieved, it becomes clear that Ripley was taking the alchemical practice of transformation between elements by "making the quadrangle round," as he put it, to the next level. This next level was known to alchemists as "squaring the circle." The process of "squaring the circle" is a sacred geometry exercise which was seen as an allegory for the process of spiritual perfection by the alchemists. As we can see from the formula in the opening epigraph of this chapter, "squaring the circle" is actually just the first in a series of four steps involving three different geometric figures. "Make of a man and woman a circle; then a quadrangle; out of this a triangle; make again a circle, and you will have the Stone of the Wise." When we transcribe a square around a circle, a triangle around the square, and another circle around all three—we arrive at an image like the following.

Before proceeding to examining these four shapes in terms of the human developmental cycle and the Mercury elemental year returns, first imagine these four shapes as indicative of four types of consciousness or awareness. Starting in the middle or interior of the mandala, imagine the inner smaller circle as corresponding with Jean Gebser's *archaic* structure.[51] Just as the circle describes infinity, with no beginning or end, humanity had at first a consciousness that was undifferentiated from its environment, such that anywhere one looked, they were surrounded by a vast sea of consciousness which was everywhere interpenetrated by the presence of eternal spirit. In terms of Ripley's four levels of perfection, this corresponds with what he termed "Origin."

According to Gebser, the next structure of awareness for humanity is the *magic* structure, which then corresponds to the square, and thus perhaps a more differentiated awareness—into four cardinal directions, four seasons, four elements etc. It is interesting that Ripley refers to the next level of perfection as "Imperfection," because this would be consistent with both the religious concept of a "fall" from an original state of grace (or wholeness symbolized by the circle) and the Hermetic image of the soul descending through the heavenly spheres during incarnation and becoming subject to the fate of corruption in the imperfect sublunary realms of the four elements. However, for the Hermeticists, we do not have to take this fate sitting down; we can and should learn to use the magic of transformation via the square of opposites to re-attain a more perfected state.

This more perfected state would be represented by the triangle, which then corresponds to Gebser's *mythical* structure. It is more than interesting then that the Christian Bible presents a mythos of a Holy Trinity as the answer to, or salvation for, a corrupted world. For the Hermeticist, this trinity is also the unity offered via the *tria prima* or three universal substances of mercury, sulphur and salt. For Ripley, this level of perfection was referred to simply as "Perfection," though in light of the terminology that follows it, we might more aptly refer to it as the first or personal perfection.

After the triangle we have another larger circle, and this then corresponds to Gebser's *mental* structure. For Ripley, the fourth state of perfection was called *plusquam perfectum*, meaning "more than perfect," or perhaps "super perfect." It is interesting then that the main products of the mental structure, namely science and western civilization, tend to see themselves as somehow improving on the inherent perfection of nature, even as they appear to have initiated a mass extinction and what some call the anthropocene, the first geological epoch to arise as a result of human activities. From a spiritual perspective, the proper understanding of this circle is the ouroboros, the serpent devouring its own tail in a cycle of eternal change. Notice also that we have returned to the same shape as the so-called "primitive" view of the undifferentiated *archaic* structure, where everything is everywhere interpenetrated by the presence of eternal spirit.

After centuries with the scientific method as the definitive collective thought-form, we now know that the very act of observing an experiment can also affect the outcome; that the linguistic labels, categories and structure we use to convey information can also actually shape the way we see, and thus experience the world; and that the various disparate economies of the world also all act together as one system. Many other recent developments in the natural sciences, such as Heisenberg's uncertainty principle, quantum mechanics, chaos theory, and, most of all, complexity theory all seem to show that a simplistic cause-effect linear view of the world is incomplete at best, and that Cartesian reductionism and the Newtonian paradigm of "hard science" are seriously limited. Put candidly, the world of simple mechanisms and simple systems is a useful but ultimately fictitious world created by science. The truth of this is nowhere easier to see than at the end of a drug company advertisement where, after being told of the potential for simple symptom relief, a laundry list of potential "side-effects" is rushed through at the end.

The more we try to reduce things into simple systems the more and more we find only complex systems of inter-relations. This idea of the ultimate truth being complex, relative and/or relational seems to echo Gebser's final structure, the *integral* structure, toward which he asserts that humanity is moving in terms of collective awareness. Just as repeated purifications bring more and more clarity to any liquor, proponents of Gebser's integral awareness feel that it brings a diaphanous or translucent quality to awareness, and is defined by the ability to see across and through *all* of the structures, simultaneously.[52] This is the clear light of the awareness granted by discovery of the quintessence or philosopher's stone. In terms of the figure above, this of course corresponds to the totality of the image, as all four shapes are taken in at once, seen and appreciated as an integrated whole. This quality and ability of "seeing through" is demonstrated by the god Hermes on the day of his birth, as he saw or imagined his invention of the lyre from the tortoise's shell, even as the poor beast was still alive.[53]

In this light, we can actually turn the oft-hurled accusations of science and the rational mental structure back upon them, and claim quite correctly that ultimately science is really only mere "pseudo-Hermeticism." By focusing exclusively on the material reality, and indeed by logical necessity denying the very existence of the spiritual, science has been involved in a centuries long tragically myopic investigation of only one of the *tria prima*, namely salt. We have completely dissected the body of the universe but without access to soul or spirit, are nevertheless unable to even imagine putting it back together again.

On the other hand, as the Hermeticist employs the reduction of the first half of the alchemical formula (*solve et coagula*) they are reducing things into *three* parts: salt, sulphur, and mercury—body, soul, and spirit. So you see, it is precisely the complexity which science has been utterly blind to for centuries upon which the alchemist relies! Since the Mercury principle of spirit is one that is constantly flowing and weaving a balance between all pairs of opposites, then we can always count on it to distribute any adjustments made to any individual part of the system into the whole. Furthermore, it is only by attending to and purifying all three aspects of self, and then re-unifying them (*coagula*) that real, lasting transformation can occur. In this way it can be said that what is really happening in alchemy or magic, in terms of Gebser's structures, is that the alchemist "re-treats" back into the deeper structures, of the triangle or square (and the shaman into the middle *archaic* circle), and then "re-turns" to the outer circle and re-establishes wholeness. Herein, I believe, is the true magical correspondence of the retrograde process, to which we have collectively been blind.

In order to apply these levels of perfection and awareness to our understanding of the planet Mercury, and his transformational dance, we can simply use the same diagram. Over the course of seventy-nine years, the basic equivalent of a human lifetime, the Mercury retrogrades will return to the same element(s) as any birth or event, a total of twelve times. If we then divide this large circle into the four shapes of our image, then we have four major developmental periods of personal perfection: Origin, Imperfection, Perfection, and Super or Collective Perfection. Each of these four periods is made of three Mercury elemental year returns. The three elemental returns of each period then activate the nigredo, albedo, rubedo sequence of transformation outlined in chapter four.

Now we have a microcosmic developmental model of the human lifetime that appears to mirror Gebser's macrocosmic model of the evolution of human consciousness. Therefore, we can enter into a correspondence with human consciousness itself, and perhaps help realize Gebser's vision of an integral consciousness for humanity. Each of us is born into a particular vintage of the Mercury elemental year. This vintage contains an answer to a basic human collective need, question, or problem. As we develop and perfect ourselves and our lives over the course of the twelve elemental year returns, we also develop and perfect our answer to the collective question or need to which we have been assigned. This in turn ultimately serves to perfect the collective awareness.

THE INNER CIRCLE: Origin

The first three returns of the Mercury elemental year represent a process of discovery and familiarization with the hidden collective currents within the personal chart and life. After the first return, we spend much of our time in school almost exclusively surrounded with those who share the same essential vintage of the Mercury elemental year. Yet even within this common stew, there are four trigons activated and three different possible pre-natal houses activated by each trigon, thus at minimum twelve different personal combinations with which to become familiar, compare, and test ourselves. The third return of this period is the first return that occurs during official adulthood, and this becomes a mandate to perfect our personal expression of our original vintage so that we can enter and become a working member of the collective.

Mercury Elemental Year Return #1: Age 6–7 Years

The first return of the Mercury retrogrades to the element of the birth year activates deep hidden potentials of contrary and alternative expressions in all three houses of the birth trigon. Children begin to notice the ways in which they are different from the mainstream, and the desire to test more diverse options in a few areas naturally arises. At the survival level, any feelings of being different could result in being cut off from the herd, which instinctively feels very dangerous. However the complete denial of differences and pushing of these alternative currents down into the personal unconscious can be just as dangerous. This is a delicate balance and subtle dance, and will remain so at least until the third return.

In most areas, children need to develop basic feelings of social competence, which involves winning approval by demonstrating specific basic competencies that are valued by society.[54] They need to be encouraged and reinforced for their initiative and industriousness in reaching these goals, while at the same time feeling free to experiment with a few alternatives in a safe way, even if just by reading, thinking or talking about others who are different. Depending on the rest of the chart and the social conditioning thus far, they may also begin to act out these differences in a few areas now, and the support of an understanding adult is crucial to maintaining healthy expressions of differences which can (and perhaps should) eventually lead to the emergence of a unique personal genius.

This personal genius is what animistic societies refer to as "medicine." Each of us has our own completely unique energetic signature. At this developmental stage, children are awakening to an experience of their own unique medicine, and their relationship to the basic self. This return is deeply connected with the alchemical principle of Sulphur, the inner fire, animal instincts and spiritual willpower that drives us through difficulties.

As adults, any feelings of social inferiority or incompetence may have their roots within this developmental period. Any urges for alternative expressions that were inhibited or denied during this period could also be damming up the flow of a typical sense of initiative to meet other, more every-day goals. During Mercury retrogrades, and especially during other elemental returns, it may be possible to go back to soothe and heal any feelings of rejection our inner child is still holding onto. Remembering, embracing, and re-activating our connections to any animals, plants, environments, images, and creative activities that we were drawn to during this stage might represent a specific "medicine" with which to accomplish this healing. The exercises at the end of chapter three are also designed to re-activate and re-energize the inner alchemical Sulphur. Take care to remember however that the energy of a seven-year-old should be carefully and consciously blended with the wisdom of our older and wiser self.

Mercury Elemental Year Return #2: Age 12-14 Years

The second return of the Mercury retrogrades to the element of the birth year releases fundamental images of the birth vintage from deep within the psyche. These images come bursting into personal awareness by way of a profound experience or series of experiences wherein this image is witnessed in others who serve as heroic exemplars of collective roles. These experiences become imprinted in the consciousness and serve as both a driving force away from normalcy and a sublime calling toward just-as-yet-being-formed collective possibilities.

This return is also co-incident with the first opposition of Saturn to its place at birth. At this age, children are becoming young adults. They are more independent, and begin to look at the future, especially in terms of career and vocational roles. The individual wants to belong to society and fit in, and so they have to begin to learn the roles they will occupy as an adult.[55] During this stage the adolescent will re-examine their identity and try out various roles to find out exactly who they are. It is during this return that individuals are particularly impressionable to any roles, which come crashing into their awareness through major collective experiences. For instance, Paul McCartney turned fourteen in 1956 when Elvis Presley was bursting into collective awareness. McCartney has been quoted as saying: "When I heard 'Heartbreak Hotel' — I thought, this is it."[56] It was only eight years later that Beatlemania had its turn to explode onto the scene, which became a huge influence on a young Tom Petty, who was turning fourteen at the time.[57]

It is precisely just such an experience of "it-ness" that can happen when the personal seed image(s) of collective identity being released into awareness meets a heroic exemplar. If these profound experiences are not forthcoming, this return could entail more searching and exploration of the lesser-known tributaries feeding into the main streams of collective consciousness. Some people may begin to more gradually synthesize the "it" image for themselves from amongst several less monumental but nonetheless personally important formative experiences.

As adults, any feelings of identity crisis or role confusion may have their roots within this developmental period. Any urges for alternative expressions of identity, which were inhibited or denied during this period, could also be damming up the flow of a normal sense of identity, belonging, and direction. During Mercury retrogrades, and especially during other elemental returns, it may be possible to go back and explore alternative identities and/or the roots of negative identities. The exercises at the end of the inferior conjunction/initiation/albedo section of chapter four, especially the vision quest ceremony, may be especially helpful to those living through or experiencing a recapitulation of this stage. Take care to remember however that the idealism of a fourteen year old should be carefully and consciously tempered by the experience of our older and wiser self.

Mercury Elemental Year Return #3: Age 19-21 Years

The third return of the Mercury retrogrades to the element of the birth year activates a deep instinctive need to bring to fruition and perfection the roles and identities discovered during and since the previous elemental return. The major developmental need just emerging at this age is for true intimacy, both in terms of relationships with others as well as with oneself.[58] Meanwhile, the identity crisis which emerged during the previous return may still be partially or even mostly unresolved, such that at this time many people can seemingly face a choice of "settling," for a role or identity which gains them intimacy but at the cost of their most authentic individuality. Others may choose to "settle" in terms of intimacy, allowing a safe or easy partnership to become solidified because it enables them to pursue their quest for personal perfection. Perhaps because this is the first elemental return experienced with the agency of an adult, another way this desire to perfect the personal expression of the birth element can manifest is through a volatile breaking away from anything seen as less than completely genuine and an abandonment of half-measures for full immersion into and fidelity towards one's personally defined path and unique medicine. For instance, Bob Dylan left behind rural Minnesota for New York City and Jimi Hendrix stopped playing in backing bands and went to London to form his own band during this return. Later, each completed a vaunted trilogy of albums before the following return.

For most, the perfection of this origin story will be a slow and gradual process which culminates in them taking up a position of authority in the world around the time of the next Mercury elemental year return, which is also around the same time as the first Saturn return. For some whose star blazes especially bright, it seems the perfection of the origin period serves as a lifetime accomplishment. Perhaps it is because the next Mercury elemental return ushers in a period which involves the arduous process of dealing with the imperfections of the outside world of the collective, some eternally youthful souls choose to exit our drama after achieving the perfection this return inaugurates. This could be one correspondence with the so-called "27 club." For those who remain, we must surrender whatever level of perfection of our origin story we have managed to achieve back to the process of transformation at the next return.

As post–Saturn return adults, any feelings of identity crisis, role confusion, loneliness, or isolation may have their roots within this developmental period. Any of the ways we were forced to settle before our process of origin perfection was allowed to culminate could also be damming up the flow of a normal sense of identity and intimacy. During Mercury retrogrades, and especially during other elemental returns, it may be possible to go back, explore, and perfect alternative endings to our origin story. The exercise for creation of the personal philosopher's stone at the end of the morning elongation/return/rubedo section of chapter four, may be especially helpful to those living through or experiencing a recapitulation of this stage. Take care to remember however that the charisma of a twenty-one-year-old should be carefully and consciously blended with the wisdom of our older and wiser self, with proper boundaries and selectiveness being maintained.

THE INNER SQUARE: Imperfection

The second set of three returns of the Mercury elemental year to the birth element, together, define a period of development which is largely about the process of entering into relationship and dealing with the imperfections of the social collective world. The volatile idealism of youth must forge a working relationship with the fixed nature of tradition and the status quo. These battles leave scars and necessitate compromises, constantly refining the ego through the crucible of personal choice. By the end of this period, the individual has reached mid-life, and the realization sets in that only so much more can be accomplished in a lifetime, requiring even further refinement of choice as well as renewed focus.

Mercury Elemental Year Return #4: Age 26–28 Years

The fourth return of the Mercury retrogrades to the element of the birth year activates a deep instinctive need to bring the perfected or idealized form of the personal origin story into the collective, thereby seeding the world with the personal spirit, just as the personal spirit was seeded by the collective during the previous period. This collision of volatilized ideas and ideals with the relatively stubborn fixity of the collective status quo necessitates a reaction, wherein both substances are changed. However, except in unusual cases, like individual waves breaking over a rocky shoreline, the rate of change to the collective is often nearly imperceptible.

In this way we can see that the basic developmental need activated here is very similar to that of the first return of the previous period. Just as children need to be encouraged and reinforced in order to develop a sense of personal industry, these adults need to be similarly welcomed and rewarded in order to develop the capacity for a truly transformative collective contribution. This is where the concept of "selling out" arises. At this age many people are starting a family, and can no longer afford the lean lifestyle of youthful idealism. So the prospect of a good paycheck is highly motivating and can lead to another possibility of "settling"—this time for a collective role that fulfills one's needs for intimacy but somewhat neglects the need for personal development. However, physical rewards are simply not enough. The soul must be fed as well. If not allowed to burn away collective falsity and continue to fully develop their own unique personal genius and talents, their own "medicine," people will eventually develop a sense of lack of motivation, low self-esteem, and lethargy. For instance, it was at twenty-seven that Henry David Thoreau went off to live alone for two years in a cabin at Walden Pond and just after this return that Kurt Vonnegut, Jr. left his job at General Electric to become a full-time writer.

For most, the natural vigor of youth generally pushes these feelings aside at this time, but usually only to re-surface later. For middle-aged adults struggling with any feelings of lack of authenticity, meaning or purpose in their lives, they may look back and find those feelings have their roots within this developmental period. Loss of contact with one's own unique medicine can only be remedied by first burning away the layers of falsity that have built up over it. This is why it is important and necessary to undergo the calcining fires of the nigredo on a semi-regular basis. The longer we put them off, the hotter the burn we will need in the end.

During Mercury retrogrades, and especially during other elemental returns, it may be possible to go back, explore, and reconnect with our personal medicine in such a way as to renew its ability to burn space in the collective for something new. The exercises for creation of the personal philosopher's stone at the end of the evening elongation/separation/nigredo section of chapter four, may be especially helpful to those living through or experiencing a recapitulation of this stage. Take care to remember however that the idealism of a twenty-seven-year-old must now be more carefully and consciously applied with the wisdom of an older adult, like the focused flame of a blowtorch, if it is to burn more deeply and effectively through layers of collective dogma.

Mercury Elemental Year Return #5: Age 32–34 Years

Like the middle return of the previous period, the fifth return of the Mercury retrogrades to the element of the birth year once again releases fundamental images of the birth vintage from deep within the psyche. However, rather than projecting these images onto a heroic exemplar, the challenge here is for integration and becoming a personal embodiment of them. This comes from deepening one's relationship to self.

For most people, the identity crisis that began in the previous period has been resolved. Generally speaking, people score consistently on personality indices after the age of thirty, and so the primary developmental need now becomes that of intimacy. This return is not only the Mercury elemental year recurring in the same element as the birth year, but also near the same degrees. Furthermore, this happens at a time when there are also major Venus and Mars returns occurring together for the first time in life. This is a time when intimacy can be achieved on many levels. Most importantly the sacred marriage to one's inner opposite can be sanctified.

The seed of this inner union produces a divine child. The "itness" experienced from without during the previous period (specifically around the second Mercury elemental year return at ages 12–14) is now experienced for the first time within. The bright light of the albedo illuminates the collective purpose of the individual, resulting in a creative surge. Branching out, trying, and perfecting new forms or genres within the chosen field can bring increased attention and popularity to one's work. It may even be that the first glimpse of the personal masterpiece arrives during this return and the years that follow. For instance, Thomas Jefferson was thirty-three when he wrote the Declaration of Independence.

As middle-aged adults, any feelings of isolation, particularly from oneself or ones creative muse may have their roots within this developmental period. Any urges for deeper personal explorations that were inhibited or denied during this period could also be damming up the flow of a normal sense of personal intimacy and self-knowledge. During Mercury retrogrades, and especially during other elemental returns, it may be possible to go back and more deeply explore and merge with one's inner opposite. The exercises at the end of the inferior conjunction/initiation/albedo section of chapter four, especially the vision quest ceremony, may be especially helpful to those living through or experiencing a recapitulation of this stage. Take care to remember however that the spiritual purity of a thirty two year old should be carefully and consciously tempered by the practical experiences of our older and wiser self.

Mercury Elemental Year Return #6: Age 38-40 Years

The sixth return of the Mercury retrogrades to the element of the birth year activates a deep instinctive need to bring to fruition and perfection the collective contribution of the individual. The major developmental need just emerging at this age is for generativity – establishing, guiding, and perfecting our contribution to the collective and the next generations.[59] The existential drive to make one's life count for something larger than oneself begins to take precedence in the person's awareness.

Jung asserted that the first half of life is about expanding into the world, and James Hollis has framed this as a developmental need for answering the question: "what does the world want of me?"[60] This return, being the last Mercury elemental return in the first half of the typical span of a human life, is when one becomes consumed in and with perfecting the answer to that question. For most, the perfection of this collective story, toward the betterment of society or guiding future generations, will be a slow and gradual process that lasts through the next period of three returns. So this period could represent the first stage of the person answering this world need and making an impact on the collective. For those whose star blazes especially bright, this return could mark a high point and a unique accomplishment that benefits the collective and upon which they can build their reputation or continue to work for some time. For instance, Nietzsche began to write *Thus Spoke Zarathustra* when he was around thirty-eight years old. In any event, the earlier challenge to develop personal intimacy now extends or evolves into the need for a kind of public or collective intimacy, where service to one's family legacy, work or profession, community or social causes takes on a central importance.

As post mid-life adults, any feelings of isolation or despair may have their roots within this developmental period. Any ways we were forced to settle before our origin perfection was then allowed to deliver its gifts to the world could also be damming up the flow of a sense of intimacy and generativity. During Mercury retrogrades, and especially during other elemental returns, it may be possible to go back, explore, and perfect alternative endings to our world story.

The exercise for creation of the personal philosopher's stone at the end of the morning elongation/return/rubedo section of chapter four, may be especially helpful to those living through or experiencing a recapitulation of this stage. Take care to remember however that the drive and ambitions of a thirty-eight-year old should be carefully and consciously blended with the wisdom of our older and wiser self, maintaining proper balance between needs of the world and the needs of the soul.

THE OUTER TRIANGLE: Personal Perfection

During the second half of a typical human life span, the first three returns of the Mercury elemental year to the birth element represent a process of turning away from the previous mode of meeting a world need, and instead bringing the attention back to oneself and one's own spiritual perfection. Jung saw the second half of life as about expanding into the inner dimensions of the personal and collective unconscious, and Hollis frames this as a developmental need for answering the question: "what does the soul want of me?"[61] Rather than trying to find and develop our identity, place, and contribution to the world, the second half of life is about finding and developing our unique soul essence. Jung called this lifelong process *individuation*, and in terms of Jungian mechanics the second half of life necessitates a movement from persona and ego development, our revealed social self, to shadow and anima/us development, our hidden interior spiritual selves.[62] Only by developing the "inferior" or unexplored parts of ourselves, those that were largely neglected in our quest to answer a world need in the first half of life, can we become more whole. As Hollis puts it: "We are not here to fit in, be well balanced, or provide exempla for others. We are here to be eccentric, different, perhaps strange, perhaps merely to add our small piece, our little clunky, chunky selves, to the great mosaic of being. As the gods intended, we are here to become more and more ourselves."[63] Of course, the particular god who presides over this transition, from the answering of the world question to the answering of the soul question, is the god who rules over *all* transitions and transitional spaces—Hermes.⁶⁴ And so it is that we may feel, see and get to know the spirit of Hermes moving through our lives at the time of mid-life, perhaps more than any other.

Mercury Elemental Year Return #7: Age 45-47 Years

The seventh return of the Mercury retrogrades to the element of the birth year activates a deep instinctive need to turn within and explore the frontiers of uncharted inner territories. Dissatisfaction with the fruits of worldly successes leads to an existential crisis wherein one wonders exactly what can be the purpose of such enormous struggle. It may take the rest of the lifetime to sufficiently answer this question, and an instinctive sense of this burden leads to cutting away of attachments to worldly pursuits. What worldliness remains must be pursued for its own sake, or for one's own sake.

The transition from the garden of our origin story to the imperfections of the world story coincided with the first Saturn return, but the transition from worldliness to soulfulness of mid-life happens after the second Saturn opposition. The revelatory nature of the opposition as a mirror remains, but rather than an innocent fourteen-year-old encountering a reflection of their "it-ness" from without, the world-hardened veteran now realizes everything that child missed out on. This unlived life is the shadow complex and underneath the dark tragic feelings of loss lie a goldmine of developmental potential. This return is the most precise of the life thus far, with Mercury repeating the triple alignments not just in the same element, but nearly the exact degrees of the birth year. This can lead to a significant release of energy and potential. Ideally, the eureka moment of discovering these untapped resources leads to a desire for personal spiritual completion, refinement, and perfection.

A great example of the pioneering possibilities that can be discovered around this return is the publication of *Sidereus Nuncius* (Starry Messenger) by Galileo in 1610. Both Galileo's refinement of the popular spyglass into a proper scientific instrument, his subsequent observation and documentation of never-before-seen phenomena that would change the world forever, and his reaction to these are all archetypal representations of the possibilities and quality of this period. Galileo's mistake was to continue to pursue his worldliness and egocentric focus in the face of such a divine revelation. With typical fiery bombast (he was born in the Mercury elemental year of fire), he became consumed with zealotry, and his lack of humility eventually condemned him to house arrest for the remainder of his life. Such a forced introspection is the inevitable result of neglect for the inner process of refinement that must accompany any outer refinements. As without, so within.

To avoid such a fate ourselves, we must remember we are not serving to answer a world need any longer, but are instead in service to the soul, and thus the divine. As Hollis puts it: "to ask what the soul wants of me is to submit to what the gods wish...to serve the gods, not the ego, not the tribe, not one's parents, not one's prior picture, is to transform."[65] As starry or divine messengers we must remember that we are only a vehicle and so must take pains to purify and refine that vehicle to better receive and reveal the sacred hierophanies with which we have been entrusted.

As adults in the second half of life, any feelings of stagnation, world-weariness, absurdity or meaninglessness may have their roots in this developmental period. Any ways we were unable to devote time to our own inner experiences could also be damming up the flow of a typical sense of spiritual purpose. During Mercury retrogrades, and especially during other elemental returns, it may be possible to go back, explore and seek to refine and perfect previously neglected aspects of ourselves. The exercises at the end of the nigredo and albedo sections of chapter four may be helpful to those living through or experiencing a recapitulation of this stage. Take care to remember however that the intense need to turn within of a forty-six-year-old should be carefully and consciously blended with the wisdom of our older and wiser self, with proper balance between our mundane life and soul needs being maintained.

Mercury Elemental Year Return #8: Age 52-54 Years

As the middle return of this period, the eighth return of the Mercury retrogrades to the element of the birth year has the potential to activate and release a shower of light from within the deepest interiors of the individual. If the fires of the nigredo phase were allowed to burn away worldly falsity during and since the previous return, then the deepest brilliance of the unique personal medicine can be accessed and released into the collective at this time. The albedo represents a successful marriage with the inner opposite, and with the genius of both sides of the personality joined, this can result in works considered as personal masterpiece. This may of course result in tremendous worldly success, but this is only possible as a result of deep inner communion with the soul. This release of the personal soul essence into the world can result in tremendous changes to collective awareness. For instance, Beethoven's ninth symphony was debuted at this time. Though he was completely deaf, Beethoven's ninth is widely acknowledged as one of the supreme masterpieces of the western tradition. Adam Smith was fifty-three at the time of publication of *The Wealth of Nations*.

During the first half of life, we are taught and conditioned to find and keep our place within the pre-determined social structures. However, the second half of life is about the need, and indeed the responsibility, to shake off the chains of conformity to those structures and reveal our own unique personal genius to ourselves, and the world. As storyteller, author, and scholar Michael Meade puts it: "There's an African proverb: 'When death finds you, may it find you alive.' Alive means living your own damn life, not the life that your parents wanted, or the life some cultural group or political party wanted, but the life that your own soul wants to live. That's the way to evaluate whether you are an authentic person or not."[66]

During this return, if we are living authentically, the deep soulful truth of our own unique personal medicine, genius, and story should begin to naturally shine forth in unmistakable ways. This is similar to the deeper meaning of the Chinese term Kung Fu. Much more than simply a particular style of martial art, Kung Fu means accomplishment or excellence arrived at by great effort of time and energy. The deeper spiritual meaning is "inner truth" or "inmost sincerity" and so Kung Fu can be displayed even in the most seemingly menial tasks, such as sweeping a floor. Thus even a janitor can have great Kung Fu if it is a personally meaningful choice for them to inhabit that role and they do so with soulfulness. Indeed, hexagram sixty-one of the *I Ching*, tells us that Kung Fu moves even pigs and fish.[67] This means that when a person is so completely faithful to their unique personal medicine and genius as to remain unperturbed in hard to endure situations, this state has the same subtle power to penetrate and move everything around them that the alchemists attribute to the philosopher's stone. Thus, by practicing our own Kung Fu, we can help heal the world by simply and truly inhabiting our own soul.

To the extent that we are still living according to someone else's story, no matter how great a story it may be, we may have trouble contacting and unlocking the deepest connections to our own soul. Inauthenticity is not always easy to see or identify in ourselves, but it can often show up in the "shoulds" and "musts" we project onto others. If we are busy telling others how they should or must live, then chances are good that we are still somewhat chained to an inauthentic self. General feelings of dissatisfaction, stagnation, lack of meaning or purpose can also indicate that we need to reactivate the purgatory fires of nigredo in order burn off impurities and release the albedo, the pure white light of our own soul essence. During Mercury retrogrades, and especially during other elemental returns, it may be possible to go back, explore, and seek to refine and perfect previously neglected aspects of ourselves. Both the exercises at the end of the nigredo and albedo sections of chapter four may be helpful to those living through or experiencing a recapitulation of this stage. Another tool which Jung himself relied on heavily during his own second half of life journey into soulfulness is called "active imagination," which is a way to dialogue with the unconscious, including re-activating symbols from dreams and other spiritual states, in order to mine them more deeply for their personal soulful meanings.[68]

Mercury Elemental Year Return #9: Age 58–60 Years

As the final return of this period, the ninth return of the Mercury elemental year has the potential to bring to completion and fruition the personal perfection, merging the vivifying powers of soulfulness with the outer world vision and accomplishments of spirit. The result is the personal philosopher's stone, which represents the seed encapsulation of everything learned thus far in life. The ninth Mercury elemental year return is roughly coincident with the fifth Jupiter return and the second Saturn return, making this a most significant rite of passage into elderhood. The personal achievements made at this time are meant to serve as a concentration of the life wisdom, as well as becoming an impetus and vehicle for the social or collective perfection of the following period.

An excellent example of the overall process of these returns is Clara Barton's founding of the American Red Cross during this ninth return. Barton's natal horoscope is considered very mercurial in that Mercury in Sagittarius is about to cross the threshold of visibility; this happens when a planet is approximately fifteen degrees from the Sun.[69] In Hellenistic astrology this condition is called "phasis" meaning "an appearance" and was seen to saturate the chart and life with the significations of the planet. Indeed Barton led quite a Mercurial life, having been a schoolteacher at seventeen and later becoming the first woman to receive a substantial clerkship in the federal government (at the US Patent Office) at a salary equal to a man's during the first half of her life.

Barton was born in the Mercury elemental year of water, so it is no surprise she eventually found her way into the caring profession of nursing. It is probable she had Aries rising, and this puts her prenatal inferior conjunction in Scorpio in the eighth place and her post-natal inferior conjunction in Pisces in the twelfth place, activating the Trigon of Dynasty.[70] It is quite fitting to the conjunction in Scorpio/eighth that Barton served as a nurse on the front lines of the American Civil War just after her sixth Mercury elemental year return. The period leading up to her seventh return found her running the Office of Missing Soldiers, where she and her assistants wrote nearly fifty thousand replies to inquiries about missing soldiers and helped find, identify, and properly bury tens of thousands of soldiers—fitting to a post-natal conjunction in Pisces/twelfth.

During the time of her seventh return, Barton achieved widespread recognition by delivering lectures around the country about her war experiences. After her work in the Franco-Prussian War, she returned to the United States and inaugurated a movement to gain recognition for the International Committee of the Red Cross by the United States government at the time of her eighth return. This movement finally bore fruit at her ninth return, when she became President of the American branch of the society, which held its first official meeting May 21, 1881 when Barton was sixty years old. She remained President, working for collective perfection throughout her next three returns, until she was forced to resign in 1904, at the age of eighty-three.

Not everyone is born to make the kind of contribution that Barton made to the collective. Nonetheless we do each have our own personal perfection, however humble it may seem to us at the time, that brings the spiritual essence of our life's work to fruition. This seed encapsulation may be hard to identify if our developmental needs at earlier stages were not fully met or resolved, but it is still there. General feelings of dissatisfaction, stagnation, lack of meaning or purpose at this stage may indicate the presence of rigidity in the emotional, mental, familial or social patterns that are hindering the creative expression.[71] Gently loosening any rigidity through patient exploration of creative exercises may help to bring the expansive view of one's life story into focus. If this is not sufficient it may be necessary to re-activate the purgatory fires of nigredo in order burn through the rigidity and release the insights of the albedo, before the rubedo process of distilling the stone of one's personal perfection can happen. Somehow there is a seed that we are meant to find, shepherd, and later bring out and plant in the collective. During Mercury retrogrades, and especially during other elemental returns, it may be possible to go back, explore, and seek to refine and perfect our view of ourselves and our life. All the exercises in chapter four may be helpful to those living through or experiencing a recapitulation of this stage.

THE OUTER CIRCLE: Collective Perfection

The next three returns of the Mercury elemental year represent a process of bringing the personal perfection into the collective and becoming a catalyst for social changes through the integral power of one's own perfected being. The humanist psychologist Abraham Maslow is famous for creating a model of human motivation based not on pathology, but on developmental growth through meeting basic universal needs, which he organized as a hierarchy. At first, Maslow expressed the need for self-actualization as being the highest. Maslow also saw that the seeking and realization of one's personal potential leads to "peak experiences"—profoundly transcendent moments of love, happiness and understanding, during which a person feels more whole, alive, at one with the world, and more aware of higher principles such as truth, justice, and harmony. Because they are characterized by wholeness and oneness, it could be that these peak experiences are glimpses of Gebser's integral structure of consciousness in practice. Later in his career, Maslow realized that self-actualization is not the highest need a human being can fulfill. Like many others, he saw that transcendent experiences activate still higher needs. Ultimately, the self only finds its actualization in giving itself to some higher goal outside oneself, in altruism and spirituality. Maslow's highest level, above self-actualization, then became transcendence needs, which involves helping others to achieve their own self-actualization.[72]

In this way, Maslow helped to clarify the basic nature and purpose of being an elder. Because of their many achievements, it can be easy to see the role of the elder as simply to enjoy retirement, and passing on all their accumulated knowledge and possessions to the next generations. But the truly wise elder offers not just knowledge, experience, and sagacious counsel but the gift of *integrity*—the ability to see across and through the many layers of life they have lived and the many levels of consciousness they have witnessed. This gift can be of great service in facilitating the shift in collective consciousness toward the new integral awareness that Gebser saw us moving towards. By sharing their own personal integrity with the world and facilitating the self-actualization of others, the true elder can reach the state of *plusquam perfectum*, becoming "more-than-perfect" by facilitating collective perfection, or the evolution of culture and consciousness itself. As storyteller, author, and scholar Michael Meade puts it: "Awakening the unique potential in all people is especially important today, to combat the conformity that mass culture increasingly demands. True culture arises from the creative depths of one's self and one's life situation."[73]

Mercury Elemental Year Return #10: Age 65-67 Years

The tenth return of the Mercury retrogrades to the element of the birth year activates a deep instinctive need to find ways in which one can begin to serve to bring perfection to the social or collective environment. This is not about establishing one's place in the world, seeking social changes out of the need for egoic elbow room, rather this involves putting aside personal needs as much as possible and seeking to facilitate the needs of others, the needs of the collective, and ultimately the needs of the epoch itself. More than likely this will involve a shift in focus from the sign/house/element of the pre-natal inferior conjunction to the sign/house/element of the post-natal inferior conjunction. For some this shift will have already happened at an earlier age, in which case it will become more pronounced and the native may even be able to begin to effectively access the transformative currents of the third member of the birth trigon at this time.

A great example of the possibilities inherent in this return is Dr. Francis Townsend, a sixty-seven-year-old physician from Long Beach, California, who in early 1933 suggested that the government pay $200 per month for Americans sixty years old or older who agreed not to work and to spend the money right away. Townsend published his plan in a local newspaper, as a kind of extended letter to the editor. Poverty among the elderly was a major social problem of the time and the letter generated a swift and massive response that led to the formation of an organization and the development of a formal plan. The plan was then published as a pamphlet and distributed throughout the country. President Franklin Roosevelt's response was to pass the Social Security Act, which included much less generous benefits. Townsend continued his activism, and his plan helped to induce amendments to the Social Security Act in 1939, which upgraded benefits and the plan indirectly spurred the augmentation of Social Security again in 1950.

Of course, this ability to tune in to the collective and act as a personal agent for the collective will is not the only way one can use the energies of this first return of the final period. One could just as easily contribute to catalyzing and focusing the self-actualization needs of other individuals. If the natural processes of life do not seem to present opportunities for either, then one could choose to activate the transformational energies by practicing the give-away ceremony at the end of the nigredo section of chapter four. In fact, some see the energy of the give-away as crucial to becoming an elder, so that even if someone is involved in social causes dear to their heart and contributing to the self-actualization needs of other individuals, the practice of further give-away can still be beneficial.[74] In particular, give-away medicine may be crucial in helping to release the fear of dying. A generation of older people who use the medicine of give-away to courageously become elders could truly change the world.

Mercury Elemental Year Return #11: Age 71–73 Years

The eleventh return of the Mercury retrogrades to the element of the birth year activates a deep outpouring of energy by which one can serve to actualize and bring to perfection the needs of the social, collective environment, or the needs of the epoch in which they live. As the albedo phase of this period, the eleventh return releases energy that has been bound up in other pursuits, like making a living. Freedom gained from answering to the world now pours forth the answer to the soul question.

An excellent example of the possibilities inherent in this return is Michael Meade's new book, *The Genius Myth*.[75] This book comes after decades of experience working in the men's movement and taking on difficult world problems, such as disaffected youth, gang violence, returning veterans, and bigotry. Meade's mission has developed from courageously facing the troubles found in the dark margins of our society into the wisdom of a compassionate elder, finding solutions to the more universal soul problems that underlie these manifestations. He contends that what is missing from our world is the recognition of genius, creativity, and imagination inherent in all humanity. The sense of an innate genius imbedded in each human soul is an ancient and valuable concept, and Meade proposes that "the accumulated vitality of many lives being lived more fully can restore and re-story the world" and that "when connected to an innate sense of 'dharma' or genuine service to the world, human genius can become a light shining forth in a time of darkness."[76] Meade uses the story of Romanian artist Ion Barladeanu as an example. Barladeanu had been impoverished for over twenty years following the fall of the Communist government in Bucharest, living as a tramp and often surviving on scraps. Even this destitute life was not enough to quell his genius however; he made art from magazine and newspaper scraps he found in the refuse. Discovered in 2008, well into his sixties, Barladeanu's collages have now made him famous and his works have been exhibited alongside artists such as Andy Warhol and Marcel Duchamp.

Meade believes that "genius fosters genius," and that as more and more people discover their innate genius, we can collectively resurrect the world from even the most dire circumstances.[77]

Meade also notes that genius is often troubling to society and similarly that genius often arises from times of trouble. Just as Campbell divided mythic themes between the social dogmas of the village orthodoxy versus the forest adventure of the hero/ine, Meade notes that society creates fixed patterns for people to follow, but genius is unique and individual. In order to find our personal genius we have to "get off the map" and into the forests and wilderness of our own souls.[78] Elders are in a peculiarly special place to accomplish this. Because their societal responsibilities are relaxed, they can perhaps most afford to wander the paths of soul. If one is having trouble navigating this wilderness, the exercises at the end of the albedo section of chapter four are designed to help reveal the natural light of the soul. According to Jung, kindling this soul light may be the whole reason we are here in the first place. "As far as we can discern, the sole purpose of human existence is to kindle a light in the darkness of mere being. It may even be assumed that just as the unconscious affects us, so the increase in our consciousness affects the unconscious."[79]

Mercury Elemental Year Return #12: Age 78-80 Years

The twelfth return of the Mercury retrogrades to the element of the birth year activates a deep, instinctive need to summarize, concentrate, and bring to perfection the personal life story, especially as it represents an answer to the needs of the social, collective environment or the needs of the epoch in which one lives. As the rubedo phase of this period, the twelfth return helps us to bring together all the seemingly disparate pieces and threads of the life into a unified whole, giving a palpable sense of some larger organizing force at work.

C. G. Jung authored his autobiographical *Memories, Dreams, Reflections* during this return, and I have previously mentioned (in the rubedo section of chapter four) that his archetype of the Self was envisioned as just such an invisible but overarching organizing force, residing outside the personal ego in the collective unconscious. Another great example of the possibilities inherent in this return is *The Power of Myth*, a series of six one-hour conversations between mythologist Joseph Campbell and journalist Bill Moyers, which were recorded during the final two summers of Campbell's life. These conversations serve to bring together Campbell's diverse life work in a way that is spontaneous and yet whole and unified. In the final conversation, "Masks of Eternity," Campbell comments on an essay by the philosopher Schopenhauer, using it as an example for what he sees as a universal theme of a hidden, structuring force behind the individual life that is itself part of a larger symphony.[80] In fact, near the end of his own life Schopenhauer commented on this sense of an overarching unity and purpose to the individual life, as if it were a play or novel authored by an outside force or genius. Schopenhauer regarded this "transcendent fatalism," as he called it, as being not only commonly experienced but historically accepted dogma and used an appeal to another respected authority as part of his argument for its existence.[81]

This overarching and interpenetrating type of vision or perception is reminiscent of Gebser's integral structure of consciousness. And while it seems (for now, at least) that it may be most naturally or fully developed during this period of life, and especially this return, it need not be the only time that this awareness is available to us. Rather, I imagine that at this stage of life each of the peak experiences from the previous periods could be seen as merely the temporal expression of the same essential guiding force which classically was thought to accompany and protect every person, whatever name we wish to call it by, whether the Greek *daimon*, Roman *genius*, or Christian *guardian angel*.

Moreover, just as the Romans also saw the power of genius to preside over places (*genius loci*) as well as people, the German term *Zeitgeist*, meaning "spirit of the age," suggests that because an individual artist is a product of his or her time, they therefore bring the culture of that time to any given work of art. Perhaps this is what humanistic astrologer Alexander Ruperti meant when he said, during an interview after this last return: "the essential task of astrologers should be to interpret human life in terms of the needs of the epoch."[82] Because Ruperti goes on to say that "what the world asks of us at any given time and the real destiny of any person, whatever their superficial life may be, is always to resolve some basic human problem," he seems to suggest that there is a correspondence between the deeper soul purpose of an individual and the deeper soul purpose of their time.

Another way of saying this might be that the *Anima Mundi* or World Soul tasks each of us with formulating an individual answer to an essential collective human question according to the nature of our time. Because the Mercury elemental year is a quality of the broader time in which a person is born (rather than the individual moment defined by the horoscope or chart) then this technique can be classified as meeting Ruperti's definition of the essential task of astrologers. Each of the elemental returns can then be seen to re-activate that essential birth vintage in such a way as to reveal deeper and deeper layers of both the personal destiny and basic human problem which needs answered. This final return then represents a chance to formally present our "closing arguments" for having answered that collective problem or need in our own individual way.

SECTION III
Appendix: Mercury Elemental Year Ephemeris

MERCURY ELEMENTAL YEAR EPHEMERIS 1925-2050

THE TABLES IN this section display five crucial bits of information about the Mercury retrograde cycle for any 16-24 month period during the century and a quarter studied. Displayed you will find:

1. **Element**: the primary element in which Mercury performs his backward trickster medicine dance

2. **Max E ES**: the date and zodiacal longitude (sign/degree) of Mercury's maximum elongation from the sun as Evening Star (his highest appearance *above* the western horizon prior to retrograde motion)

3. **Conjunction**: the date and zodiacal longitude (sign/degree) of Mercury's inferior conjunction with the sun. This happens at the mid-point of the retrograde period (approx 10-11 days after station retrograde and 10-11 days before station direct). Conjunction occurs after Mercury disappears *below* the horizon, hence while Mercury is invisible. Conjunction is also known as *cazimi*, or "in the heart of the sun," and symbolizes a secret meeting of the messenger Hermes in the Throne Room of the King. The zodiacal signs in which these secret meetings occur define the Mercury Elemental Year.

4. **Max E MS**: the date and zodiacal longitude (sign/degree) of Mercury's maximum elongation from the Sun as Morning Star (his highest appearance *above* the eastern horizon following retrograde motion).

5. **Stars**: the stars and constellations located in these degrees often reveal profound archetypal themes and provide important clues to further meanings, experiences, and magical applications.[83] Stars marked with an asterisk (*) are either particularly bright, close to the ecliptic, Royal Stars of Persia, and/or Behenian stars.[84]

To Find your Personal Mercury Elemental Year

In the following tables, look up the most recent conjunction *prior to* your birth. This happened at some time during your last four months of gestation in the womb, and the energetic reverberations of this most important alignment remained and were imprinted within your unconscious mind at birth. After locating the most recent conjunction *prior to* your birth, note the element to which it belongs and also the zodiacal sign and degree.

The personal Mercury Elemental Year represents a fundamental frame of reference or contextual vessel for your personal horoscope or birth chart. No matter what sign Mercury is found in at the moment of your birth, the natal placement should be seen to operate within the broader context of the Mercury Elemental Year, which represents an unconscious preference for one of four modes of symbol making.

After this primary elemental mode is grasped, sometimes the stellar symbolism is also quite worthy of investigation, often yielding deeper or even more specific clues to the magical manifestations of these elemental currents. Star input can be more subtle, nuanced, or generalized to the constellation, especially with dimmer stars, or those located at a distance from the ecliptic, that is, those without an asterisk.

Inferior conjunctions of Mercury occur about every 116 days, or close to four months apart. As a general rule, at any moment in time the most recent *previous* conjunction to birth defines the most important of the three signs of the Mercury Elemental Year. At some point in life, most often during the second half, the post-natal conjunction will become more important, both to individual development and collective contributions.

Occasionally, about twenty percent of people will be born during the four-month period between the last conjunction in one element and the first conjunction in the next element. The prenatal conjunction is still the most important of these, though eventually, some will be able to identify with the collective currents of *both* elements and miraculously manage to alchemically blend them together.

The short period between the first and/or last maximum elongation and conjunction in an elemental period could be expressed more easily as an alchemical blend of two elements—especially in rarer instances when the first and/or last maximum elongation and conjunction of a period occur in different signs/elements.

Mercury Elemental Year Ephemeris 1925-2050

Dates from **12/11/1925** to **03/12/1927** belong the Mercury Elemental Years of Fire

ELEMENT △	Max E Eve Star Above W Horizon	Conjunction Below Horizon (other planets also in conjunction)	Max E Morn Star Above E Horizon	Stars in these degrees & (Constellation)
1) FIRE	11/22/1925 Sagittarius 23	12/11/1925 Sagittarius 20	12/31/1925 Sagittarius 18	Ras Alhague (Ophiucus)
2) FIRE	3/16/1926 Aries 12	03/31/1926 Aries 10	04/27/1926 Aries 11	Alderamin (Cepheus)
3) FIRE	07/10/1926 Leo 15	08/07/1926 Leo 15	08/25/1926 Leo 14	Dubhe (Ursa Major)
4) FIRE	11/04/1926 Sagittarius 6	11/25/1926 Sagittarius 3 (Venus)	12/13/1926 Sagittarius 1	Dschubba (Scorpius)

Dates from **03/13/1927** to **02/05/1929** belong the Mercury Elemental Years of Water

ELEMENT ▽	Max E Eve Star Above W Horizon	Conjunction Below Horizon (other planets also in conjunction)	Max E Morn Star Above E Horizon	Stars in these degrees & (Constellation)
1) WATER	2/25/1927 Pisces 25	3/13/1927 Pisces 23	4/09/1927 Pisces 22	Markab (Pegasus)
2) WATER	6/22/1927 Cancer 26	7/19/1927 Cancer 27	8/08/1927 Cancer 27	Procyon* (Canis Minor)
3) WATER	10/18/1927 Scorpio 19	11/10/1927 Scorpio 17 (Mars)	11/26/1927 Scorpio 15	Zuben Eschamali* (Scales)
4) WATER	02/08/1928 Pisces 8	02/24/1928 Pisces 5	03/22/1928 Pisces 5	Deneb Adige* (Cygnus)

5) WATER	6/02/1928 Cancer 6	06/29/1928 Cancer 8 (Venus)	7/21/1928 Cancer 9	Alhena (Gemini)
6) WATER	09/29/1928 Scorpio 3	10/24/1928 Scorpio 1	11/09/1928 Libra 28	

Dates from **02/06/1929** to **05/19/1930** belong the Mercury Elemental Years of Air

ELEMENT A	Max E Eve Star Above W Horizon	Conjunction Below Horizon (other planets also in conjunction)	Max E Morn Star Above E Horizon	Stars in these degrees & (Constellation)
1) AIR	01/22/1929 Aquarius 21	02/06/1929 Aquarius 18	03/04/1929 Aquarius 17	
2) AIR	05/15/1929 Gemini 17	06/09/1929 Gemini 19	07/03/1929 Gemini 21	Bellatrix* (Orion)
3) AIR	9/12/1929 Libra 17	10/08/1929 Libra 15	10/23/1929 Libra 12	Kraz (Corvus)
4) AIR	01/05/1930 Aquarius 5	01/21/1930 Aquarius 2	02/16/1930 Aquarius 1	ALTAIR* (Aquila)

Dates from **05/20/1930** to **12/20/1931** belong the Mercury Elemental Years of Earth

ELEMENT ▽	Max E Eve Star Above W Horizon	Conjunction Below Horizon (other planets also in conjunction)	Max E Morn Star Above E Horizon	Stars in these degrees & (Constellation)
1) EARTH	04/27/1930 Taurus 28	05/20/1930 Taurus 29	06/15/1930 Gemini 1	Alcyone* (Pleiades)
2) EARTH	08/25/1930 Virgo 30	09/21/1930 Virgo 29	10/07/1930 Virgo 26	
3) EARTH	12/20/1930 Capricorn 18	01/05/1931 Capricorn 15 (Saturn)	01/28/1931 Capricorn 14	Vega* (Lyra)

4) EARTH	04/10/1931 Taurus 9	04/30/1931 Taurus 10	05/27/1931 Taurus 11	
5) EARTH	08/08/1931 Virgo 13	09/04/1931 Virgo 12 (Venus)	09/20/1931 Virgo 10	Zosma (Leo)

Dates from **12/21/1931** to **11/17/1933** belong the Mercury Elemental Years of Fire

ELEMENT △	Max E Eve Star Above W Horizon	Conjunction Below Horizon (other planets also in conjunction)	Max E Morn Star Above E Horizon	Stars in these degrees & (Constellation)
1) FIRE	12/02/1931 Capricorn 2*	12/21/1931 Sagittarius 29 (Winter Solstice)	01/11/1932 Sagittarius 27	Galactic Center (Milky Way)
2) FIRE	03/23/1932 Aries 22	04/10/1932 Aries 21 (Uranus)	05/08/1932 Aries 22	Baten Kaitos (Cetus)
3) FIRE	07/20/1932 Leo 25	08/17/1932 Leo 25	09/03/1932 Leo 23	Alphard (Hydra)
4) FIRE	11/14/1932 Sagittarius 16	12/04/1932 Sagittarius 13	12/23/1932 Sagittarius 11	Ras Algethi (Hercules)
5) FIRE	03/06/1933 Aries 4	03/23/1933 Aries 3 (Vernal Point)	04/19/1933 Aries 3	Deneb Kaitos (Cetus)
6) FIRE	07/02/1933 Leo 7	07/30/1933 Leo 7	08/17/1933 Leo 7	Asselus Australis (Cancer)

Dates from **11/18/1933** to **02/16/1935** belong the Mercury Elemental Years of Water

ELEMENT ▽	Max E Eve Star Above W Horizon	Conjunction Below Horizon (other planets also in conjunction)	Max E Morn Star Above E Horizon	Stars in these degrees & (Constellation)
1) WATER	10/28/1933 Scorpio 29	11/19/1933 Scorpio 27	12/06/1933 Scorpio 24	Toliman* (Centaurus)
2) WATER	02/18/1934 Pisces 18	03/06/1934 Pisces 15	04/02/1934 Pisces 15	Achernar (Eradinus) Ankaa (Phoenix)

3) WATER	06/14/1934 Cancer 18	07/11/1934 Cancer 19 (Moon)	07/31/1934 Cancer 19	Castor* (Gemini)
4) WATER	10/10/1934 Scorpio 13	11/02/1934 Scorpio 11 (Venus & Jupiter)	11/19/1934 Scorpio 8	Alphecca* (Corona Borealis)

Births from **02/17/1935** thru **01/13/1937** belong the Mercury Elemental Years of Air

ELEMENT △	Max E Eve Star Above W Horizon	Conjunction Below Horizon (other planets also in conjunction)	Max E Morn Star Above E Horizon	Stars in these degrees & (Constellation)
1) AIR	02/01/1935 Pisces 1	02/17/1935 Aquarius 28 (Saturn)	03/15/1935 Aquarius 27	Enif (Pegasus)
2) AIR	05/26/1935 Gemini 28	06/21/1935 Gemini 30 (Summer Solstice)	07/14/1935 Cancer 1	Betelgeuse* (Orion)
3) AIR	9/23/1935 Libra 26	10/18/1935 Libra 24	09/02/1935 Libra 22	Spica (Virgo)* Arcturus (Bootes)*
4) AIR	01/16/1936 Aquarius 15	01/31/1936 Aquarius 11	02/26/1936 Aquarius 10	
5) AIR	05/07/1936 Gemini 9	05/31/1936 Gemini 10 (Mars)	06/26/1936 Gemini 13	Aldebaran* (Taurus)
6) AIR	9/04/1936 Libra 10	09/30/1936 Libra 8	10/16/1936 Libra 6	Diadem (Coma Berenices) Porrima, Vindemiatrix (Virgo)

Births from **01/14/1937** thru **12/13/1938** belong the Mercury Elemental Years of Earth

ELEMENT ▽	Max E Eve Star Above W Horizon	Conjunction Below Horizon (other planets also in conjunction)	Max E Morn Star Above E Horizon	Stars in these degrees & (Constellation)
1) EARTH	12/29/1936 Capricorn 28	01/14/1937 Capricorn 25	02/07/1937 Capricorn 24	

2) EARTH	04/20/1937 Taurus 20	05/11/1937 Taurus 21	06/06/1937 Taurus 23	
3) EARTH	08/17/1937 Virgo 23	09/14/1937 Virgo 22 (Neptune)	09/30/1937 Virgo 19	Alkes (Crater)
4) EARTH	12/12/1937 Capricorn 12	12/29/1937 Capricorn 8	01/20/1938 Capricorn 7	Facies M22 (Sagittarius)
5) EARTH	04/02/1938 Taurus 2	04/21/1938 Taurus 2	05/19/1938 Taurus 3	
6) EARTH	07/31/1938 Virgo 6	08/28/1938 Virgo 5	09/13/1938 Virgo 3	Thuban (Draco)

Births from **12/14/1938** thru **03/14/1940** belong the Mercury Elemental Years of Fire

ELEMENT △	Max E Eve Star Above W Horizon	Conjunction Below Horizon (other planets also in conjunction)	Max E Morn Star Above E Horizon	Stars in these degrees & (Constellation)
1) FIRE	11/25/1938 Sagittarius 25	12/14/1938 Sagittarius 22	01/03/1939 Sagittarius 20	Ras Alhague (Ophiucus)
2) FIRE	3/16/1939 Aries 14	04/03/1939 Aries 13	05/01/1939 Aries 14	Alpheratz (Andromeda)
3) FIRE	07/13/1939 Leo 18	08/10/1939 Leo 18	08/29/1939 Leo 17	
4) FIRE	11/07/1939 Sagittarius 8	11/28/1939 Sagittarius 6	12/17/1939 Sagittarius 4	Antares* (Scorpius)

Births from **03/15/1940** thru **02/08/1942** belong the Mercury Elemental Years of Water

ELEMENT ▽	Max E Eve Star Above W Horizon	Conjunction Below Horizon (other planets also in conjunction)	Max E Morn Star Above E Horizon	Stars in these degrees & (Constellation)
1) WATER	02/28/1940 Pisces 27	03/15/1940 Pisces 25	04/12/1940 Pisces 25	

2) WATER	06/24/1940 Cancer 29	07/21/1940 Cancer 30	08/10/1940 Cancer 30	
3) WATER	10/20/1940 Scorpio 22	11/11/1940 Scorpio 20	11/28/1940 Scorpio 17	Zuben Eschamali* (Scales)
4) WATER	02/10/1941 Pisces 11	02/26/1941 Pisces 8 (Moon)	03/25/1941 Pisces 7	Deneb Adige* (Cygnus)
5) WATER	06/06/1941 Cancer 10	07/02/1941 Cancer 11	07/24/1941 Cancer 12	Sirius* (Canis Major)
6) WATER	10/02/1941 Scorpio 5	10/26/1941 Scorpio 4	11/11/1941 Scorpio 1	

Births from **02/09/1942** thru **01/07/1944** belong the Mercury Elemental Years of Air

ELEMENT △	Max E Eve Star Above W Horizon	Conjunction Below Horizon (other planets also in conjunction)	Max E Morn Star Above E Horizon	Stars in these degrees & (Constellation)
1) AIR	01/25/1942 Aquarius 24	02/09/1942 Aquarius 21	03/07/1942 Aquarius 20	Sadalsuud* (Aquarius) Deneb Algedi* (Capricornus)
2) AIR	05/18/1942 Gemini 20	06/12/1942 Gemini 22	07/06/1942 Gemini 24	El Nath* (Taurus)
3) AIR	09/15/1942 Libra 19	10/10/1942 Libra 18 (Mars)	10/26/1942 Libra 15	
4) AIR	01/08/1943 Aquarius 8	01/24/1943 Aquarius 4	02/18/1943 Aquarius 3	
5) AIR	04/30/1943 Gemini 1	05/23/1943 Gemini 2 (Uranus)	06/18/1943 Gemini 4	Mirfak (Perseus)
6) AIR	08/28/1943 Libra 3	09/24/1943 Libra 1 (Neptune)	10/10/1943 Virgo 29	

Births from **01/08/1944** thru **04/12/1945** belong the Mercury Elemental Years of Earth

ELEMENT ▽	Max E Eve Star Above W Horizon	Conjunction Below Horizon (other planets also in conjunction)	Max E Morn Star Above E Horizon	Stars in these degrees & (Constellation)
1) EARTH	12/23/1943 Capricorn 21	01/08/1944 Capricorn 18	01/31/1944 Capricorn 16	
2) EARTH	04/12/1944 Taurus 12	05/02/1944 Taurus 13	05/29/1944 Taurus 15	Menkar (Cetus)
3) EARTH	08/10/1944 Virgo 16	09/06/1944 Virgo 15 (Jupiter)	09/22/1944 Virgo 13	
4) EARTH	12/05/1944 Capricorn 5	12/22/1944 Capricorn 2	01/13/1945 Sagittarius 30	

Births from **04/13/1945** thru **11/20/1946** belong the Mercury Elemental Years of Fire

ELEMENT △	Max E Eve Star Above W Horizon	Conjunction Below Horizon (other planets also in conjunction)	Max E Morn Star Above E Horizon	Stars in these degrees & (Constellation)
1) FIRE	03/25/1945 Aries 24	04/13/1945 Aries 24 (Venus)	05/11/1945 Aries 25	
2) FIRE	07/23/1945 Leo 28	08/20/1945 Leo 28	09/06/1945 Leo 26	Regulus* (Leo)
3) FIRE	11/17/1945 Sagittarius 18	12/07/1945 Sagittarius 15	12/26/1945 Sagittarius 13	
4) FIRE	03/09/1946 Aries 7	03/26/1946 Aries 6	04/22/1946 Aries 6	
5) FIRE	07/05/1946 Leo 10	08/02/1946 Leo 10 (Pluto)	08/20/1946 Leo 9	

Births from **11/21/1946** thru **10/18/1948** belong the Mercury Elemental Years of Water

ELEMENT ▽	Max E Eve Star Above W Horizon	Conjunction Below Horizon (other planets also in conjunction)	Max E Morn Star Above E Horizon	Stars in these degrees & (Constellation)
1) WATER	10/31/1946 Sagittarius 2	11/21/1946 Scorpio 29 (Venus)	12/09/1946 Scorpio 27	Toliman* (Centaurus)
2) WATER	02/20/1947 Pisces 20	03/08/1947 Pisces 18	04/05/1947 Pisces 18	
3) WATER	06/17/1947 Cancer 21	07/14/1947 Cancer 22	08/03/1947 Cancer 22	Pollux* (Gemini)
4) WATER	10/13/1947 Scorpio 15	11/05/1947 Scorpio 13 (Chiron)	11/22/1947 Scorpio 10	Zuben Elgenubi* (Scales)
5) WATER	02/04/1948 Pisces 4	02/19/1948 Pisces 1	03/17/1948 Aquarius 30	Sadalmelek (Aquarius) Fomalhaut* (Pisces Austrinus)
6) WATER	05/28/1948 Cancer 1	06/23/1948 Cancer 3 (Venus)	07/16/1948 Cancer 4	

Births from **10/19/1948** thru **01/16/1950** belong the Mercury Elemental Years of Air

ELEMENT △	Max E Eve Star Above W Horizon	Conjunction Below Horizon (other planets also in conjunction)	Max E Morn Star Above E Horizon	Stars in these degrees & (Constellation)
1) AIR	09/25/1948 Libra 29	10/19/1948 Libra 27	11/4/1948 Libra 24	
2) AIR	01/17/1949 Aquarius 17	02/02/1949 Aquarius 14	02/27/1949 Aquarius 13	ALNAIR (Grus)
3) AIR	05/10/1949 Gemini 12	06/03/1949 Gemini 13	06/28/1949 Gemini 16	Rigel* (Orion)
4) AIR	09/07/1949 Libra 12	10/03/1949 Libra 11 (Neptune)	10/19/1949 Libra 8	Gienah* (Corvus)

Births from **01/17/1950** thru **12/15/1951** belong the Mercury Elemental Years of Earth

ELEMENT ▽	Max E Eve Star Above W Horizon	Conjunction Below Horizon (other planets also in conjunction)	Max E Morn Star Above E Horizon	Stars in these degrees & (Constellation)
1) EARTH	01/01/1950 Aquarius 1	01/17/1950 Capricorn 28	02/10/1950 Capricorn 26	
2) EARTH	04/22/1950 Taurus 23	05/14/1950 Taurus 24	06/10/1950 Taurus 26	Algol* (Perseus)
3) EARTH	08/20/1950 Virgo 25	09/17/1950 Virgo 24 (Saturn)	10/02/1950 Virgo 22	Cor Caroli (Canes Venatici)
4) EARTH	12/15/1950 Capricorn 14	01/01/1951 Capricorn 11	01/23/1951 Capricorn 9	Nunki (Sagittarius)
5) EARTH	04/05/1951 Taurus 5	04/24/1951 Taurus 5 (Mars)	05/22/1951 Taurus 6	Hamal (Aries)
6) EARTH	08/03/1951 Virgo 8	08/31/1950 Virgo 8 (Venus)	09/16/1951 Virgo 6	Alioth (Ursa Major)

Births from **12/16/1951** thru **03/17/1953** belong the Mercury Elemental Years of Fire

ELEMENT △	Max E Eve Star Above W Horizon	Conjunction Below Horizon (other planets also in conjunction)	Max E Morn Star Above E Horizon	Stars in these degrees & (Constellation)
1) FIRE	11/28/1951 Sagittarius 28	12/16/1951 Sagittarius 25	01/06/1952 Sagittarius 23	
2) FIRE	03/18/1952 Aries 17	04/05/1952 Aries 16	05/04/1952 Aries 17	
3) FIRE	07/15/1952 Leo 20	08/12/1952 Leo 20 (Pluto)	08/30/1952 Leo 19	
4) FIRE	11/09/1952 Sagittarius 11	11/30/1952 Sagittarius 9	12/18/1952 Sagittarius 6	Antares* (Scorpius)

Births from **03/18/1953** thru **07/24/1953** belong the Mercury Elemental Year of Water

ELEMENT ▽	Max E Eve Star Above W Horizon	Conjunction Below Horizon (other planets also in conjunction)	Max E Morn Star Above E Horizon	Stars in these degrees & (Constellation)
1) WATER	03/02/1953 Pisces 30	03/18/1953 Pisces 28 (Vernal Point)	04/15/1953 Pisces 28	Scheat (Pegasus)

Births from **07/25/1953** thru **11/13/1953** belong the Mercury Elemental Year of Fire

ELEMENT △	Max E Eve Star Above W Horizon	Conjunction Below Horizon (other planets also in conjunction)	Max E Morn Star Above E Horizon	Stars in these degrees & (Constellation)
1) FIRE	06/27/1953 Leo 2	07/25/1953 Leo 3	08/13/1953 Leo 2	

Births from **11/14/1953** thru **02/11/1955** belong the Mercury Elemental Years of Water

ELEMENT ▽	Max E Eve Star Above W Horizon	Conjunction Below Horizon (other planets also in conjunction)	Max E Morn Star Above E Horizon	Stars in these degrees & (Constellation)
1) WATER	10/23/1953 Scorpio 25	11/14/1953 Scorpio 23	12/01/1953 Scorpio 20	
2) WATER	02/13/1954 Pisces 13	03/01/1954 Pisces 11 (Venus)	03/28/1954 Pisces 10	Achernar (Eradinus)
3) WATER	06/09/1954 Cancer 13	07/05/1954 Cancer 14 (Jupiter)	07/27/1954 Cancer 15	Canopus* (Argo Navis)
4) WATER	10/05/1954 Scorpio 8	10/29/1954 Scorpio 6 (Saturn)	11/14/1954 Scorpio 4	

Births from **02/12/1955** thru **01/09/1957** belong the Mercury Elemental Years of Air

ELEMENT △	Max E Eve Star Above W Horizon	Conjunction Below Horizon (other planets also in conjunction)	Max E Morn Star Above E Horizon	Stars in these degrees & (Constellation)
1) AIR	01/28/1955 Aquarius 27	02/12/1955 Aquarius 24	03/10/1955 Aquarius 23	Sadalsuud* (Aquarius) Deneb Algedi* (Capricornus)
2) AIR	05/21/1955 Gemini 23	06/16/1955 Gemini 25	07/09/1955 Gemini 27	Saiph (Orion)
3) AIR	09/18/1955 Libra 22	10/13/1955 Libra 20	10/29/1955 Libra 17	Spica* (Virgo) Arcturus* (Bootes)
4) AIR	01/11/1956 Aquarius 10	01/27/1956 Aquarius 7 (Chiron)	02/21/1956 Aquarius 6	
5) AIR	05/02/1956 Gemini 4	05/25/1956 Gemini 5	06/20/1956 Gemini 7	Prima Hyadum (Hyades/Taurus)
6) AIR	08/30/1956 Libra 5	09/26/1956 Libra 4	10/11/1956 Libra 2	

Births from **01/10/1957** thru **04/15/1958** belong the Mercury Elemental Years of Earth

ELEMENT ▽	Max E Eve Star Above W Horizon	Conjunction Below Horizon (other planets also in conjunction)	Max E Morn Star Above E Horizon	Stars in these degrees & (Constellation)
1) EARTH	12/24/1956 Capricorn 24	01/10/1957 Capricorn 21	02/02/1957 Capricorn 19	Peacock (Pavo)
2) EARTH	04/15/1957 Taurus 15	05/05/1957 Taurus 16 (Venus)	06/01/1957 Taurus 18	
3) EARTH	08/13/1957 Virgo 19	09/09/1957 Virgo 17 (Mars)	09/25/1957 Virgo 15	
4) EARTH	12/08/1957 Capricorn 7	12/25/1957 Capricorn 4	01/16/1958 Capricorn 3	

Births from **04/16/1958** thru **03/09/1960** belong the Mercury Elemental Years of Fire

ELEMENT △	Max E Eve Star Above W Horizon	Conjunction Below Horizon (other planets also in conjunction)	Max E Morn Star Above E Horizon	Stars in these degrees & (Constellation)
1) FIRE	03/28/1958 Aries 27	04/16/1958 Aries 27	05/14/1958 Aries 28	Alrisha (Pisces)
2) FIRE	07/26/1958 Virgo 1	08/23/1958 Leo 30 (Pluto)	09/09/1958 Leo 29	Regulus* (Leo)
3) FIRE	11/20/1958 Sagittarius 21	12/09/1958 Sagittarius 18	12/29/1958 Sagittarius 16	
4) FIRE	03/12/1959 Aries 10	03/29/1959 Aries 8	04/26/1959 Aries 9	
5) FIRE	07/08/1959 Leo 13	08/05/1959 Leo 13 (Uranus)	08/23/1959 Leo 12	Dubhe (Ursa Major)
6) FIRE	11/03/1959 Sagittarius 4	11/24/1959 Sagittarius 2	12/12/1959 Scorpio 30	Dschubba (Scorpius)

Births from **03/10/1960** thru **10/21/1961** belong the Mercury Elemental Years of Water

ELEMENT ▽	Max E Eve Star Above W Horizon	Conjunction Below Horizon (other planets also in conjunction)	Max E Morn Star Above E Horizon	Stars in these degrees & (Constellation)
1) WATER	02/23/1960 Pisces 23	03/10/1960 Pisces 21	04/07/1960 Pisces 21	Markab (Pegasus)
2) WATER	06/19/1960 Cancer 24	07/16/1960 Cancer 25	08/05/1960 Cancer 25	Procyon* (Canis Minor)
3) WATER	10/15/1960 Scorpio 18	11/07/1960 Scorpio 16 (Neptune)	11/24/1960 Scorpio 13	Zuben Elgenubi* (Scales)
4) WATER	02/06/1961 Pisces 6	02/21/1961 Pisces 4 (Chiron)	03/20/1960 Pisces 3	Deneb Adige* (Cygnus)

5) WATER	06/01/1961 Cancer 5	06/27/1961 Cancer 6	07/19/1961 Cancer 7	

Births from **10/22/1961** thru **01/19/1963** belong the Mercury Elemental Years of Air

ELEMENT △	Max E Eve Star Above W Horizon	Conjunction Below Horizon (other planets also in conjunction)	Max E Morn Star Above E Horizon	Stars in these degrees & (Constellation)
1) AIR	09/28/1961 Scorpio 1	10/22/1961 Libra 30	11/07/1961 Libra 27	
2) AIR	01/20/1962 Aquarius 20	02/05/1962 Aquarius 17 (Venus & Jupiter)	03/02/1962 Aquarius 15	Sualocin (Delphinus)
3) AIR	05/13/1962 Gemini 15	06/07/1962 Gemini 17	07/01/1962 Gemini 19	Rigel* (Orion)
4) AIR	09/10/1962 Libra 15	10/06/1962 Libra 13	10/22/1962 Libra 11	

Births from **01/20/1963** to **12/17/1964** belong the Mercury Elemental Years of Earth

ELEMENT ▽	Max E Eve Star Above W Horizon	Conjunction Below Horizon (other planets also in conjunction)	Max E Morn Star Above E Horizon	Stars in these degrees & (Constellation)
1) EARTH	01/04/1963 Aquarius 3	01/20/1963 Capricorn 30	02/13/1963 Capricorn 29	ALTAIR* (Aquila)
2) EARTH	04/25/1963 Taurus 26	05/17/1963 Taurus 27	06/13/1963 Taurus 29	Algol* (Perseus) Alcyone* (Pleiades/ Taurus)
3) EARTH	08/23/1963 Virgo 28	09/19/1963 Virgo 27 (Venus)	10/05/1963 Virgo 25	Alkaid* (Ursa Major)
4) EARTH	12/18/1963 Capricorn 17	01/04/1964 Capricorn 14	01/26/1964 Capricorn 12	Vega* (Lyra)

5) EARTH	04/07/1964 Taurus 8	04/27/1964 Taurus 8 (Jupiter)	05/24/1964 Taurus 9	Hamal (Aries)
6) EARTH	08/04/1964 Virgo 11	09/02/1964 Virgo 10 (Uranus & Pluto)	09/18/1964 Virgo 8	Zosma (Leo)

Births from **12/18/1964** to **11/16/1966** belong the Mercury Elemental Years of Fire

ELEMENT △	Max E Eve Star Above W Horizon	Conjunction Below Horizon (other planets also in conjunction)	Max E Morn Star Above E Horizon	Stars in these degrees & (Constellation)
1) FIRE	11/30/1964 Sagittarius 30	12/18/1964 Sagittarius 27	01/08/1965 Sagittarius 26	Galactic Center (Milky Way)
2) FIRE	03/21/1965 Aries 20	04/08/1965 Aries 19 (Venus)	05/06/1965 Aries 20	Baten Kaitos (Cetus)
3) FIRE	07/18/1965 Leo 23	08/15/1965 Leo 23	09/01/1965 Leo 22	Alphard (Hydra)
4) FIRE	11/13/1965 Sagittarius 14	12/02/1965 Sagittarius 11	12/21/1965 Sagittarius 9	Antares* (Scorpius)
5) FIRE	03/04/1966 Aries 3	03/21/1966 Aries 1 (Mars & Vernal point)	04/18/1966 Aries 1	Deneb Kaitos (Cetus)
6) FIRE	06/30/1966 Leo 5	07/28/1966 Leo 6	08/16/1966 Leo 5	

Births from **11/17/1966** to **02/14/1968** belong the Mercury Elemental Years of Water

ELEMENT ▽	Max E Eve Star Above W Horizon	Conjunction Below Horizon (other planets also in conjunction)	Max E Morn Star Above E Horizon	Stars in these degrees & (Constellation)
1) WATER	10/26/1966 Scorpio 27	11/17/1966 Scorpio 25 (Venus & Neptune)	12/04/1966 Scorpio 23	

2) WATER	02/16/1967 Pisces 16	03/04/1967 Pisces 14	03/31/1967 Pisces 13	Achernar (Eradinus) Ankaa (Phoenix)
3) WATER	06/12/1967 Cancer 16	07/09/1967 Cancer 17	07/30/1967 Cancer 18	Castor* (Gemini)
4) WATER	10/08/1967 Scorpio 11	11/01/1967 Scorpio 9	11/17/1967 Scorpio 6	Alphecca* (Corona Borealis)

Births from **02/15/1968** to **01/12/1970** belong the Mercury Elemental Years of Air

ELEMENT	Max E Eve Star Above W Horizon	Conjunction Below Horizon (other planets also in conjunction)	Max E Morn Star Above E Horizon	Stars in these degrees & (Constellation)
1) AIR	01/30/1968 Aquarius 29	02/15/1968 Aquarius 27	03/12/1968 Aquarius 25	Gienah (Cygnus)
2) AIR	05/23/1968 Gemini 26	06/18/1968 Gemini 28 (Venus & Mars)	07/11/1968 Gemini 29	Betelgeuse* (Orion) Polaris (Ursa Minor)
3) AIR	09/20/1968 Libra 25	10/15/1968 Libra 23	10/31/1968 Libra 20	Spica* (Virgo) Arcturus* (Bootes)
4) AIR	01/13/1969 Aquarius 13	01/29/1969 Aquarius 10	02/23/1969 Aquarius 9	
5) AIR	05/05/1969 Gemini 7	05/29/1969 Gemini 8	06/23/1969 Gemini 10	Aldebaran* (Taurus)
6) AIR	09/02/1969 Libra 8	09/29/1969 Libra 7 (Uranus)	10/14/1969 Libra 4	Diadem (Coma Berenices)

Births from **01/13/1970** to **04/18/1971** belong the Mercury Elemental Years of Earth

ELEMENT	Max E Eve Star Above W Horizon	Conjunction Below Horizon (other planets also in conjunction)	Max E Morn Star Above E Horizon	Stars in these degrees & (Constellation)
1) EARTH	12/27/1969 Capricorn 27	01/13/1970 Capricorn 23 (Venus)	02/05/1970 Capricorn 22	Peacock (Pavo)

2) EARTH	04/18/1970 Taurus 18	05/09/1970 Taurus 19 (Saturn)	06/05/1970 Taurus 21	
3) EARTH	08/16/1970 Virgo 21	09/12/1970 Virgo 20	09/28/1970 Virgo 18	Denabola (Leo)
4) EARTH	12/10/1970 Capricorn 10	12/28/1970 Capricorn 7 (Moon)	01/19/1971 Capricorn 5	Facies (Sagittarius)

Births from **04/19/1971** to **08/25/1971** belong the Mercury Elemental Year of Fire

ELEMENT △	Max E Eve Star Above W Horizon	Conjunction Below Horizon (other planets also in conjunction)	Max E Morn Star Above E Horizon	Stars in these degrees & (Constellation)
1) FIRE	03/31/1971 Aries 30	04/19/1971 Aries 30	05/17/1971 Taurus 1	Mirach (Andromeda)

Births from **08/26/1971** to **12/11/1971** belong the Mercury Elemental Years of Earth

ELEMENT ▽	Max E Eve Star Above W Horizon	Conjunction Below Horizon (other planets also in conjunction)	Max E Morn Star Above E Horizon	Stars in these degrees & (Constellation)
1) EARTH	07/29/1971 Virgo 3	08/26/1971 Virgo 3 (Venus)	09/12/1971 Virgo 2	

Births from **12/12/1971** to **03/12/1973** belong the Mercury Elemental Years of Fire

ELEMENT △	Max E Eve Star Above W Horizon	Conjunction Below Horizon (other planets also in conjunction)	Max E Morn Star Above E Horizon	Stars in these degrees & (Constellation)
1) FIRE	11/23/1971 Sagittarius 23	12/12/1971 Sagittarius 21 (Jupiter)	01/01/1972 Sagittarius 19	Ras Alhague (Ophiucus)
2) FIRE	03/14/1972 Aries 13	03/31/1972 Aries 11 (Chiron)	04/28/1972 Aries 12	Alderamin (Cepheus)

3) FIRE	07/10/1972 Leo 16	08/07/1972 Leo 16	08/25/1972 Leo 15	Dubhe (Ursa Major)
4) FIRE	11/05/1972 Sagittarius 7	11/25/1972 Sagittarius 5 (Neptune)	12/14/1972 Sagittarius 2	

Births from **03/13/1973** to **02/07/1975** belong the Mercury Elemental Years of Water

ELEMENT ▽	Max E Eve Star Above W Horizon	Conjunction Below Horizon (other planets also in conjunction)	Max E Morn Star Above E Horizon	Stars in these degrees & (Constellation)
1) WATER	02/25/1973 Pisces 26	03/13/1973 Pisces 24	04/10/1973 Pisces 24	Markab (Pegasus)
2) WATER	06/22/1973 Cancer 27	07/20/1973 Cancer 28	08/08/1973 Cancer 28	Procyon (Canis minor)
3) WATER	10/18/1973 Scorpio 20	11/10/1973 Scorpio 18	11/27/1973 Scorpio 16	Zuben Eschamali* (Scale)
4) WATER	02/09/1974 Pisces 9	02/24/1974 Pisces 6	03/23/1974 Pisces 6	Deneb Adige* (Cygnus)
5) WATER	06/04/1974 Cancer 8	06/30/1974 Cancer 9 (Saturn)	07/22/1974 Cancer 10	Alhena (Gemini)
6) WATER	09/30/1974 Scorpio 4	10/25/1974 Scorpio 2	11/10/1974 Libra 30*	

Births from **02/08/1975** to **05/19/1976** belong the Mercury Elemental Years of Air

ELEMENT △	Max E Eve Star Above W Horizon	Conjunction Below Horizon (other planets also in conjunction)	Max E Morn Star Above E Horizon	Stars in these degrees & (Constellation)
1) AIR	01/23/1975 Aquarius 23	02/08/1975 Aquarius 19	03/05/1975 Aquarius 18	
2) AIR	05/16/1975 Gemini 18	06/10/1975 Gemini 20	07/04/1975 Gemini 22	Bellatrix* (Orion) Capella* (Auriga)

Mercury Elemental Year Ephemeris

3) AIR	09/13/1975 Libra 18	10/09/1975 Libra 16 (Pluto)	10/24/1975 Libra 13	Kraz (Corvus)
4) AIR	01/07/1976 Aquarius 6	01/23/1976 Aquarius 3	02/16/1976 Aquarius 2	ALTAIR* (Aquila)

Births from **05/20/1976** to **12/20/1977** belong the Mercury Elemental Years of Earth

ELEMENT ▽	Max E Eve Star Above W Horizon	Conjunction Below Horizon (other planets also in conjunction)	Max E Morn Star Above E Horizon	Stars in these degrees & (Constellation)
1) EARTH	04/27/1976 Taurus 29	05/20/1976 Taurus 30	06/15/1976 Gemini 2	Alcyone* (Pleiades/ Taurus)
2) EARTH	08/26/1976 Libra 1	09/21/1976 Virgo 30	10/07/1976 Virgo 27	
3) EARTH	12/20/1976 Capricorn 20	01/06/1977 Capricorn 16	01/28/1977 Capricorn 15	Vega* (Lyra)
4) EARTH	04/10/1977 Taurus 11	04/30/1977 Taurus 11	05/27/1977 Taurus 13	
5) EARTH	08/08/1977 Virgo 14	09/05/1977 Virgo 13	09/21/1977 Virgo 11	Zosma (Leo)

Births from **12/21/1977** to **11/18/1979** belong the Mercury Elemental Years of Fire

ELEMENT △	Max E Eve Star Above W Horizon	Conjunction Below Horizon (other planets also in conjunction)	Max E Morn Star Above E Horizon	Stars in these degrees & (Constellation)
1) FIRE	12/03/1977 Capricorn 3	12/21/1977 Sagittarius 30 (Winter Solstice)	01/11/1978 Sagittarius 28	Galactic Center (Milky Way)
2) FIRE	03/24/1978 Aries 23	04/11/1978 Aries 22	05/09/1978 Aries 23	Baten Kaitos (Cetus)
3) FIRE	07/21/1978 Leo 26	08/18/1978 Leo 26	09/04/1978 Leo 25	Alphard (Hydra)

4) FIRE	11/15/1978 Sagittarius 16	12/05/1978 Sagittarius 14 (Neptune)	12/24/1978 Sagittarius 12	Ras Algethi (Hercules)
5) FIRE	03/07/1979 Aries 5	03/24/1979 Aries 4	04/21/1979 Aries 4	Deneb Kaitos (Cetus)
6) FIRE	07/03/1979 Leo 8	07/31/1979 Leo 8	08/29/1979 Leo 8	Assellus Australis (Cancer)

Births from **11/19/1979** to **02/16/1981** belong the Mercury Elemental Years of Water

ELEMENT ▽	Max E Eve Star Above W Horizon	Conjunction Below Horizon (other planets also in conjunction)	Max E Morn Star Above E Horizon	Stars in these degrees & (Constellation)
1) WATER	10/29/1979 Scorpio 30	11/19/1979 Scorpio 28 (Uranus)	12/07/1979 Scorpio 25	Toliman* (Centaurus)
2) WATER	02/19/1980 Pisces 19	03/06/1980 Pisces 16	04/02/1980 Pisces 16	
3) WATER	06/14/1980 Cancer 19	07/11/1980 Cancer 20	08/01/1980 Cancer 20	Castor* (Gemini)
4) WATER	10/10/1980 Scorpio 14	11/03/1980 Scorpio 12	11/19/1980 Scorpio 9	Alphecca* (Corona Borealis)

Births from **02/17/1981** to **06/20/1981** belong the Mercury Elemental Years of Air

ELEMENT △	Max E Eve Star Above W Horizon	Conjunction Below Horizon (other planets also in conjunction)	Max E Morn Star Above E Horizon	Stars in these degrees & (Constellation)
1) AIR	02/01/1981 Pisces 2	02/17/1981 Aquarius 29	03/15/1981 Aquarius 28	

Births from **06/21/1981** to **10/17/1981** belong the Mercury Elemental Years of Water

ELEMENT ▽	Max E Eve Star Above W Horizon	Conjunction Below Horizon (other planets also in conjunction)	Max E Morn Star Above E Horizon	Stars in these degrees & (Constellation)
1) WATER	05/27/1981 Gemini 29	06/21/1981 Cancer 1 (Summer Solstice)	07/14/1981 Cancer 3	Betelgeuse* (Orion)

Births from **10/18/1981** to **01/14/1983** belong the Mercury Elemental Years of Air

ELEMENT △	Max E Eve Star Above W Horizon	Conjunction Below Horizon (other planets also in conjunction)	Max E Morn Star Above E Horizon	Stars in these degrees & (Constellation)
1) AIR	09/23/1981 Libra 27	10/18/1981 Libra 25 (Pluto)	11/02/1981 Libra 23	Spica* (Virgo) Arcturus* (Bootes)
2) AIR	01/16/1982 Aquarius 16	01/31/1982 Aquarius 12	02/26/1982 Aquarius 11	
3) AIR	05/08/1982 Gemini 10	06/01/1982 Gemini 11	06/26/1982 Gemini 14	Aldebaran* (Taurus)
4) AIR	09/05/1982 Libra 11	10/02/1982 Libra 9	10/17/1982 Libra 7	Porrima, Vindemiatrix (Virgo)

Births from **01/15/1983** to **12/13/1984** belong the Mercury Elemental Years of Earth

ELEMENT ▽	Max E Eve Star Above W Horizon	Conjunction Below Horizon (other planets also in conjunction)	Max E Morn Star Above E Horizon	Stars in these degrees & (Constellation)
1) EARTH	12/30/1982 Capricorn 29	01/15/1983 Capricorn 26	02/08/1983 Capricorn 25	
2) EARTH	04/21/1983 Taurus 21	05/12/1983 Taurus 22 (Moon & Mars)	06/08/1983 Taurus 24	
3) EARTH	08/19/1983 Virgo 24	09/15/1983 Virgo 23	10/01/1983 Virgo 21	Alkes (Crater)

4) EARTH	12/13/1983 Capricorn 13	12/31/1983 Capricorn 10	01/22/1984 Capricorn 8	
5) EARTH	04/02/1984 Taurus 3	04/21/1984 Taurus 3	05/19/1984 Taurus 4	
6) EARTH	07/31/1984 Virgo 7	08/28/1984 Virgo 6	09/13/1984 Virgo 4	Thuban (Draco) Alioth (Ursa Major)

Births from **12/14/1984** to **03/15/1986** belong the Mercury Elemental Years of Fire

ELEMENT △	Max E Eve Star Above W Horizon	Conjunction Below Horizon (other planets also in conjunction)	Max E Morn Star Above E Horizon	Stars in these degrees & (Constellation)
1) FIRE	11/25/1984 Sagittarius 26	12/14/1984 Sagittarius 23	01/03/1985 Sagittarius 21	
2) FIRE	03/16/1985 Aries 15	04/03/1985 Aries 14 (Venus)	05/01/1985 Aries 15	Alpheratz (Andromeda)
3) FIRE	07/13/1985 Leo 19	08/10/1985 Leo 19 (Mars)	08/28/1985 Leo 18	
4) FIRE	11/08/1985 Sagittarius 10	11/28/1985 Sagittarius 7 (Saturn)	12/17/1985 Sagittarius 5	Antares* (Scorpius)

Births from **03/16/1986** to **07/22/1986** belong the Mercury Elemental Year of Water

ELEMENT ▽	Max E Eve Star Above W Horizon	Conjunction Below Horizon (other planets also in conjunction)	Max E Morn Star Above E Horizon	Stars in these degrees & (Constellation)
1) WATER	02/28/1986 Pisces 28	03/16/1986 Pisces 26	04/13/1986 Pisces 26	

Births from **07/23/1986** to **11/11/1986** belong the Mercury Elemental Years of Fire

ELEMENT △	Max E Eve Star Above W Horizon	Conjunction Below Horizon (other planets also in conjunction)	Max E Morn Star Above E Horizon	Stars in these degrees & (Constellation)
1) FIRE	06/25/1986 Cancer 30	07/23/1986 Leo 1	08/11/1986 Leo 1	

Births from **11/12/1986** to **02/09/1988** belong the Mercury Elemental Years of Water

ELEMENT ▽	Max E Eve Star Above W Horizon	Conjunction Below Horizon (other planets also in conjunction)	Max E Morn Star Above E Horizon	Stars in these degrees & (Constellation)
1) WATER	10/21/1986 Scorpio 23	11/12/1986 Scorpio 21	11/29/1986 Scorpio 18	Zuben Eschamali* (Scales)
2) WATER	02/11/1987 Pisces 12	02/27/1987 Pisces 9	03/26/1987 Pisces 9	
3) WATER	06/07/1987 Cancer 11	07/03/1987 Cancer 12	07/25/1987 Cancer 13	Sirius* (Canis Major)
4) WATER	10/04/1987 Scorpio 7	10/28/1987 Scorpio 5 (Pluto)	11/13/1987 Scorpio 2	

Births from **02/10/1988** to **01/07/1990** belong the Mercury Elemental Years of Air

ELEMENT ⍍	Max E Eve Star Above W Horizon	Conjunction Below Horizon (other planets also in conjunction)	Max E Morn Star Above E Horizon	Stars in these degrees & (Constellation)
1) AIR	01/26/1988 Aquarius 25	02/10/1988 Aquarius 22	03/07/1988 Aquarius 21	Sadalsuud* (Aquarius) Deneb Algedi* (Capricornus)
2) AIR	05/19/1988 Gemini 21	06/12/1988 Gemini 23 (Venus)	07/06/1988 Gemini 25	Alnilam (Orion) Al Hecka (Taurus)
3) AIR	09/15/1988 Libra 20	10/11/1988 Libra 19	10/26/1988 Libra 16	

4) AIR	01/09/1989 Aquarius 9	01/24/1989 Aquarius 6	02/18/1989 Aquarius 4	
5) AIR	05/01/1989 Gemini 2	05/23/1989 Gemini 3	06/18/1989 Gemini 6	Prima Hyadum (Hyades/Taurus)
6) AIR	08/28/1989 Libra 3	09/24/1989 Libra 2 (Mars)	10/10/1989 Virgo 30	

Births from **01/08/1990** to **04/13/1991** belong the Mercury Elemental Years of Earth

ELEMENT ▽	Max E Eve Star Above W Horizon	Conjunction Below Horizon (other planets also in conjunction)	Max E Morn Star Above E Horizon	Stars in these degrees & (Constellation)
1) EARTH	12/23/1989 Capricorn 22	01/08/1990 Capricorn 19 (Saturn)	01/31/1990 Capricorn 17	
2) EARTH	04/13/1990 Taurus 14	05/03/1990 Taurus 14	05/30/1990 Taurus 16	Menkar (Cetus)
3) EARTH	08/11/1990 Virgo 17	09/07/1990 Virgo 16	09/23/1990 Virgo 14	
4) EARTH	12/06/1990 Capricorn 6	12/24/1990 Capricorn 3	01/14/1991 Capricorn 1	

Births from **04/14/1991** to **11/20/1992** belong the Mercury Elemental Years of Fire

ELEMENT △	Max E Eve Star Above W Horizon	Conjunction Below Horizon (other planets also in conjunction)	Max E Morn Star Above E Horizon	Stars in these degrees & (Constellation)
1) FIRE	03/27/1991 Aries 26	04/14/1991 Aries 25 (Moon)	05/12/1991 Aries 26	
2) FIRE	07/24/1991 Leo 29	08/21/1991 Leo 29 (Jupiter)	09/07/1991 Leo 27	Regulus* (Leo)
3) FIRE	11/18/1991 Sagittarius 19	12/08/1991 Sagittarius 17	12/27/1991 Sagittarius 14	

Mercury Elemental Year Ephemeris

4) FIRE	03/09/1992 Aries 8	03/26/1992 Aries 7	04/23/1992 Aries 7	
5) FIRE	07/05/1992 Leo 11	08/02/1992 Leo 11	08/21/1992 Leo 10	

Births from **11/21/1992** to **10/20/1994** belong the Mercury Elemental Years of Water

ELEMENT ▽	Max E Eve Star Above W Horizon	Conjunction Below Horizon (other planets also in conjunction)	Max E Morn Star Above E Horizon	Stars in these degrees & (Constellation)
1) WATER	10/31/1992 Sagittarius 3	11/21/1992 Scorpio 30	12/09/1992 Scorpio 28	Toliman* (Centaurus)
2) WATER	02/20/1993 Pisces 21	03/08/1993 Pisces 19	04/05/1993 Pisces 19	
3) WATER	06/17/1993 Cancer 22	07/14/1993 Cancer 23	08/04/1993 Cancer 23	Pollux* (Gemini)
4) WATER	10/13/1993 Scorpio 16	11/05/1993 Scorpio 14	11/22/1993 Scorpio 11	Zuben Elgenubi* (Scales)
5) WATER	02/04/1994 Pisces 5	02/20/1994 Pisces 2 (Saturn)	03/18/1994 Pisces 1	Sadalmelek (Aquarius) Fomalhaut* (Pisces Austrinus)
6) WATER	05/30/1994 Cancer 3	06/25/1994 Cancer 4	07/17/1994 Cancer 5	

Births from **10/21/1994** to **01/18/1996** belong the Mercury Elemental Years of Air

ELEMENT △	Max E Eve Star Above W Horizon	Conjunction Below Horizon (other planets also in conjunction)	Max E Morn Star Above E Horizon	Stars in these degrees & (Constellation)
1) AIR	09/26/1994 Libra 30	10/21/1994 Libra 28	11/05/1994 Libra 25	

2) AIR	01/19/1995 Aquarius 18	02/03/1995 Aquarius 15	03/01/1995 Aquarius 14	ALNAIR (Grus)	
3) AIR	05/11/1995 Gemini 13	06/05/1995 Gemini 15	06/29/1995 Gemini 17	Rigel* (Orion)	
4) AIR	09/08/1995 Libra 13	10/04/1995 Libra 12 (Chiron)	10/20/1995 Libra 9	Gienah* (Corvus)	

Births from **01/19/1996** to **12/16/1997** belong the Mercury Elemental Years of Earth

ELEMENT ▽	Max E Eve Star Above W Horizon	Conjunction Below Horizon (other planets also in conjunction)	Max E Morn Star Above E Horizon	Stars in these degrees & (Constellation)
1) EARTH	01/02/1996 Aquarius 2	01/19/1996 Capricorn 29 (Neptune)	02/11/1996 Capricorn 27	
2) EARTH	04/23/1996 Taurus 24	05/14/1996 Taurus 25	06/10/1996 Taurus 27	Algol* (Perseus)
3) EARTH	08/21/1996 Virgo 27	09/17/1996 Virgo 25	10/02/1996 Virgo 23	Alkaid* (Ursa Major)
4) EARTH	12/15/1996 Capricorn 15	01/01/1997 Capricorn 12	01/24/1997 Capricorn 11	
5) EARTH	04/05/1997 Taurus 6	04/25/1997 Taurus 6 (Venus)	05/22/1997 Taurus 7	Hamal (Aries)
6) EARTH	08/03/1997 Virgo 9	08/31/1997 Virgo 9	09/16/1997 Virgo 7	Alioth (Ursa Major)

Births from **12/17/1997** to **03/18/1999** belong the Mercury Elemental Years of Fire

ELEMENT △	Max E Eve Star Above W Horizon	Conjunction Below Horizon (other planets also in conjunction)	Max E Morn Star Above E Horizon	Stars in these degrees & (Constellation)
1) FIRE	11/28/1997 Sagittarius 29	12/17/1997 Sagittarius 26	01/06/1998 Sagittarius 24	Galactic Center (Milky Way)

2) FIRE	03/19/1998 Aries 18	04/06/1998 Aries 17 (Saturn)	05/04/1998 Aries 18	
3) FIRE	07/16/1998 Leo 22	08/13/1998 Leo 22	08/31/1998 Leo 20	
4) FIRE	11/11/1998 Sagittarius 12	12/01/1998 Sagittarius 10 (Pluto)	12/20/1998 Sagittarius 8	Antares* (Scorpius)

Births from **03/19/1999** to **07/25/1999** belong the Mercury Elemental Year of Water

ELEMENT ▽	Max E Eve Star Above W Horizon	Conjunction Below Horizon (other planets also in conjunction)	Max E Morn Star Above E Horizon	Stars in these degrees & (Constellation)
1) WATER	03/03/1999 Aries 1	03/19/1999 Pisces 29 Vernal point	04/16/1999 Pisces 29	Deneb Kaitos (Cetus)

Births from **07/26/1999** to **11/14/1999** belong the Mercury Elemental Year of Fire

ELEMENT	Max E Eve Star Above W Horizon	Conjunction Below Horizon (other planets also in conjunction)	Max E Morn Star Above E Horizon	Stars in these degrees & (Constellation)
1) FIRE	06/28/1999 Leo 3	07/26/1999 Leo 4	08/14/1999 Leo 3	

Births from **11/15/1999** to **02/11/2001** belong the Mercury Elemental Years of Water

ELEMENT ▽	Max E Eve Star Above W Horizon	Conjunction Below Horizon (other planets also in conjunction)	Max E Morn Star Above E Horizon	Stars in these degrees & (Constellation)
1) WATER	10/24/1999 Scorpio 26	11/15/1999 Scorpio 24	12/02/1999 Scorpio 21	
2) WATER	02/14/2000 Pisces 14	03/01/2000 Pisces 12	03/28/2000 Pisces 11	

3) WATER	06/09/2000 Cancer 14	07/06/2000 Cancer 15 (Mars)	07/27/2000 Cancer 16	Sirius* (Canis Major) Canopus* (Argo Navis)
4) WATER	10/06/2000 Scorpio 9	10/29/2000 Scorpio 7	11/15/2000 Scorpio 5	

Births from **02/12/2001** to **01/10/2003** belong the Mercury Elemental Years of Air

ELEMENT	Max E Eve Star Above W Horizon	Conjunction Below Horizon (other planets also in conjunction)	Max E Morn Star Above E Horizon	Stars in these degrees & (Constellation)
1) AIR	01/28/2001 Aquarius 28	02/12/2001 Aquarius 25 (Uranus)	03/10/2001 Aquarius 24	Sadalsuud* (Aquarius) Deneb Algedi* (Capricornus)
2) AIR	05/22/2001 Gemini 24	06/16/2001 Gemini 26 (Jupiter)	07/09/2001 Gemini 28	Saiph (Orion)
3) AIR	09/18/2001 Libra 23	10/13/2001 Libra 21	10/29/2001 Libra 18	Spica* (Virgo) Arcturus* (Bootes)
4) AIR	01/11/2002 Aquarius 11	01/27/2002 Aquarius 8 (Venus & Neptune)	02/21/2002 Aquarius 7	
5) AIR	05/04/2002 Gemini 5	05/27/2002 Gemini 6	06/21/2002 Gemini 9	Aldebaran* (Taurus)
6) AIR	08/31/2002 Libra 6	09/27/2002 Libra 5	10/13/2002 Libra 2	

Births from **01/11/2003** to **04/15/2004** belong the Mercury Elemental Years of Earth

ELEMENT	Max E Eve Star Above W Horizon	Conjunction Below Horizon (other planets also in conjunction)	Max E Morn Star Above E Horizon	Stars in these degrees & (Constellation)
1) EARTH	12/26/2002 Capricorn 25	01/11/2003 Capricorn 22	02/03/2003 Capricorn 20	Peacock (Pavo)

2) EARTH	04/16/2003 Taurus 16	05/07/2003 Taurus 17	06/03/2003 Taurus 19	
3) EARTH	08/14/2003 Virgo 20	09/10/2003 Virgo 18 (Venus)	09/26/2003 Virgo 16	
4) EARTH	12/09/2003 Capricorn 9	12/26/2003 Capricorn 5	01/17/2004 Capricorn 4	

Births from **04/16/2004** to **08/22/2004** belong the Mercury Elemental Years of Fire

ELEMENT △	Max E Eve Star Above W Horizon	Conjunction Below Horizon (other planets also in conjunction)	Max E Morn Star Above E Horizon	Stars in these degrees & (Constellation)
1) FIRE	03/29/2004 Aries 28	04/16/2004 Aries 28	05/14/2004 Aries 29	Alrisha (Pisces)

Births from **08/23/2004** to **12/09/2004** belong the Mercury Elemental Years of Earth

ELEMENT ▽	Max E Eve Star Above W Horizon	Conjunction Below Horizon (other planets also in conjunction)	Max E Morn Star Above E Horizon	Stars in these degrees & (Constellation)
1) EARTH	07/26/2004 Virgo 2	08/23/2004 Virgo 2	09/09/2004 Leo 30	

Births from **12/10/2004** to **03/10/2006** belong the Mercury Elemental Years of Fire

ELEMENT △	Max E Eve Star Above W Horizon	Conjunction Below Horizon (other planets also in conjunction)	Max E Morn Star Above E Horizon	Stars in these degrees & (Constellation)
1) FIRE	11/20/2004 Sagittarius 22	12/10/2004 Sagittarius 19 (Pluto)	12/29/2004 Sagittarius 17	
2) FIRE	03/12/2005 Aries 11	03/29/2005 Aries 10 (Venus)	04/26/2005 Aries 10	Alderamin (Cepheus)

3) FIRE	07/08/2005 Leo 14	08/05/2005 Leo 14	08/23/2005 Leo 13	Dubhe (Ursa Major)	
4) FIRE	11/03/2005 Sagittarius 5	11/24/2005 Sagittarius 3	12/12/2005 Sagittarius 1	Dschubba (Scorpius)	

Births from **03/11/2006** to **02/05/2008** belong the Mercury Elemental Years of Water

ELEMENT ▽	Max E Eve Star Above W Horizon	Conjunction Below Horizon (other planets also in conjunction)	Max E Morn Star Above E Horizon	Stars in these degrees & (Constellation)
1) WATER	02/23/2006 Pisces 24	03/11/2006 Pisces 22	04/08/2006 Pisces 22	Markab (Pegasus)
2) WATER	06/20/2006 Cancer 25	07/18/2006 Cancer 26	08/07/2006 Cancer 26	Procyon* (Canis Minor)
3) WATER	10/16/2006 Scorpio 19	11/08/2006 Scorpio 17 (Venus & Mars)	11/25/2006 Scorpio 14	Zuben Elgenubi* Zuben Eschamali* (Scales)
4) WATER	02/07/2007 Pisces 8	02/22/2007 Pisces 5	03/21/2007 Pisces 4	Deneb Adige* (Cygnus)
5) WATER	06/02/2007 Cancer 6	06/28/2007 Cancer 7	07/20/2007 Cancer 8	Alhena (Gemini)
6) WATER	09/29/2007 Scorpio 3	10/23/2007 Scorpio 1	11/08/2007 Libra 28	

Births from **02/06/2008** to **05/17/2009** belong the Mercury Elemental Years of Air

ELEMENT △	Max E Eve Star Above W Horizon	Conjunction Below Horizon (other planets also in conjunction)	Max E Morn Star Above E Horizon	Stars in these degrees & (Constellation)
1) AIR	01/22/2008 Aquarius 21	02/06/2008 Aquarius 18 (Moon & Neptune)	03/03/2008 Aquarius 17	Sualocin (Delphinus)

2) AIR	05/14/2008 Gemini 16	06/07/2008 Gemini 18 (Venus)	07/02/2008 Gemini 20	Rigel* & Bellatrix* (Orion)
3) AIR	09/10/2008 Libra 16	10/06/2008 Libra 14	10/22/2008 Libra 12	Algorab (Corvus)
4) AIR	01/04/2009 Aquarius 4	01/20/2009 Aquarius 1 (Jupiter)	02/13/2009 Capricorn 30	ALTAIR* (Aquila)

Births from **05/18/2009** to **12/18/2010** belong the Mercury Elemental Years of Earth

ELEMENT ▽	Max E Eve Star Above W Horizon	Conjunction Below Horizon (other planets also in conjunction)	Max E Morn Star Above E Horizon	Stars in these degrees & (Constellation)
1) EARTH	04/26/2009 Taurus 27	05/18/2009 Taurus 28	06/13/2009 Taurus 30	Alcyone* (Pleiades)
2) EARTH	08/24/2009 Virgo 29	09/20/2009 Virgo 28 (Saturn)	10/05/2009 Virgo 26	Alkaid* (Ursa Major)
3) EARTH	12/18/2009 Capricorn 18	01/04/2010 Capricorn 15 (Venus)	01/27/2010 Capricorn 13	Vega* (Lyra)
4) EARTH	04/08/2010 Taurus 9	04/28/2010 Taurus 9	05/25/2010 Taurus 10	Hamal (Aries)
5) EARTH	08/06/2010 Virgo 12	09/03/2010 Virgo 11	09/19/2010 Virgo 9	Zosma (Leo)

Births from **12/19/2010** to **11/16/2012** belong the Mercury Elemental Years of Fire

ELEMENT △	Max E Eve Star Above W Horizon	Conjunction Below Horizon (other planets also in conjunction)	Max E Morn Star Above E Horizon	Stars in these degrees & (Constellation)
1) FIRE	12/01/2010 Capricorn 2	12/19/2010 Sagittarius 29 (Winter Solstice)	01/09/2011 Sagittarius 27	Acumen (Scorpius)
2) FIRE	03/22/2011 Aries 21	04/09/2011 Aries 20 (Jupiter)	05/07/2011 Aries 21	Baten Kaitos (Cetus)

3) FIRE	07/19/2011
Leo 25	08/16/2011
Leo 24 (Venus)	09/03/2011
Leo 23	
4) FIRE	11/14/2011
Sagittarius 15	12/04/2011
Sagittarius 12	12/23/2011
Sagittarius 10	Alwaid
(Draco)
5) FIRE	03/05/2012
Aries 4	03/21/2012
Aries 2 (Uranus)
(Vernal point)	04/18/2012
Aries 2	Deneb Kaitos
(Cetus)
6) FIRE	06/30/2012
Leo 6	07/28/2012
Leo 7	08/16/2012
Leo 6	Praesaepe
(Cancer)

Births from **11/17/2012 to 02/14/2014** belong the Mercury Elemental Years of Water

| ELEMENT ▽ | Max E Eve
Star Above W
Horizon | Conjunction
Below Horizon
(other planets
also in
conjunction) | Max E Morn Star
Above E Horizon | Stars
in these degrees
&
(Constellation) |
|---|---|---|---|---|
| 1) WATER | 10/26/2012
Scorpio 29 | 11/17/2012
Scorpio 26 | 12/4/2012
Scorpio 24 | Toliman*
(Centaurus) |
| 2) WATER | 02/16/2013
Pisces 17 | 03/04/2013
Pisces 15 (Venus) | 03/31/2013
Pisces 14 | Ankaa
(Phoenix) |
| 3) WATER | 06/12/2013
Cancer 17 | 07/09/2013
Cancer 18 | 07/30/2013
Cancer 19 | Castor*
(Gemini) |
| 4) WATER | 10/09/2013
Scorpio 12 | 11/01/2013
Scorpio 10
(Saturn) | 11/17/2013
Scorpio 7 | Alphecca*
(Corona Borealis) |

Dates from **02/15/2014 to 01/13/2016** belong the Mercury Elemental Years of Air

| ELEMENT △ | Max E Eve
Star Above W
Horizon
Separation/
Nigredo | Conjunction
Below Horizon
(other planets
conj)
Initiation/Albedo | Max E Morn Star
Above E Horizon
Return/Rubedo | Stars
in these degrees
&
(Constellation) |
|---|---|---|---|---|
| 1) AIR | 01/31/2014
Aquarius 30 | 02/15/2014
Aquarius 28 | 03/13/2014
Aquarius 27 | Gienah
(Cygnus) |
| 2) AIR | 05/25/2014
Gemini 27 | 06/19/2014
Gemini 29
(Summer
Solstice) | 07/12/2014
Cancer 1 | Betelgeuse*
(Orion) |

3) AIR	09/21/2014 Libra 26	10/16/2014 Libra 24 (Venus)	11/01/2104 Libra 21	Spica* (Virgo) Arcturus* (Bootes)
4) AIR	01/14/2015 Aquarius 14	01/30/2015 Aquarius 11	02/24/2015 Aquarius 10	Albali (Aquarius)
5) AIR	05/07/2015 Gemini 8	05/30/2015 Gemini 9 (Mars)	06/24/2015 Gemini 12	Aldebaran* (Taurus)
6) AIR	09/03/2015 Libra 9	09/30/2015 Libra 8	10/15/2015 Libra 5	Diadem (Coma Berenices)

Dates from **01/14/2016** to **12/11/2017** belong the Mercury Elemental Years of Earth

ELEMENT ▽	Max E Eve Star Above W Horizon Separation/ Nigredo	Conjunction Below Horizon (other planets conj) Initiation/Albedo	Max E Morn Star Above E Horizon Return/Rubedo	Stars in these degrees & (Constellation)
1) EARTH	12/29/2015 Capricorn 27	01/14/2016 Capricorn 24	02/06/2016 Capricorn 23	Peacock (Pavo)
2) EARTH	04/18/2016 Taurus 19	05/09/2016 Taurus 20	06/06/2016 Taurus 22	
3) EARTH	08/16/2016 Virgo 22	09/12/2016 Virgo 21	09/28/2016 Virgo 19	Denebola (Leo)
4) EARTH	12/11/2016 Capricorn 11	12/28/2016 Capricorn 8	01/19/2017 Capricorn 6	Facies (Sagittarius)
5) EARTH	04/01/2017 Taurus 1	04/20/2017 Taurus 1	05/17/2017 Taurus 2	Mirach (Andromeda)
6) EARTH	07/29/2017 Virgo 5	08/26/2017 Virgo 4	09/12/2017 Virgo 3	

Dates from **12/12/2017** to **03/13/2019** belong the Mercury Elemental Years of Fire

ELEMENT △	Max E Eve Star Above W Horizon Separation/ Nigredo	Conjunction Below Horizon (other planets conj) Initiation/Albedo	Max E Morn Star Above E Horizon Return/Rubedo	Stars in these degrees & (Constellation)
1) FIRE	11/24/2017 Sagittarius 25	12/12/2017 Sagittarius 22	01/01/2018 Sagittarius 20	Ras Alhague (Ophiucus)
2) FIRE	03/15/2018 Aries 14	04/01/2018 Aries 12	04/29/2018 Aries 13	Alderamin (Cepheus)
3) FIRE	07/11/2018 Leo 17	08/08/2018 Leo 17	08/26/2018 Leo 16	
4) FIRE	11/06/2018 Sagittarius 8	11/27/2018 Sagittarius 6 (Jupiter)	12/15/2018 Sagittarius 3	

Dates from **03/14/2019** to **02/07/2021** belong the Mercury Elemental Years of Water

ELEMENT ▽	Max E Eve Star Above W Horizon Separation/ Nigredo	Conjunction Below Horizon (other planets conj) Initiation/Albedo	Max E Morn Star Above E Horizon Return/Rubedo	Stars in these degrees & (Constellation)
1) WATER	02/26/2019 Pisces 27	03/14/2019 Pisces 25	04/11/2019 Pisces 25	Markab (Pegasus)
2) WATER	06/23/2019 Cancer 28	07/21/2019 Cancer 29 (Venus)	08/09/2019 Cancer 29	
3) WATER	10/19/2019 Scorpio 21	11/11/2019 Scorpio 19	11/28/2019 Scorpio 17	Zuben Eschamali* (Scales)
4) WATER	02/10/2020 Pisces 10	02/25/2020 Pisces 7	03/23/2020 Pisces 7	
5) WATER	06/04/2020 Cancer 9	06/30/2020 Cancer 10	07/22/2020 Cancer 11	Alhena (Gemini)
6) WATER	10/01/2020 Scorpio 5	10/25/2020 Scorpio 3	11/10/2020 Scorpio 1	

Dates from **02/08/2021** to **01/06/2023** belong the Mercury Elemental Years of Air

ELEMENT △	Max E Eve Star Above W Horizon Separation/ Nigredo	Conjunction Below Horizon (other planets conj) Initiation/Albedo	Max E Morn Star Above E Horizon Return/Rubedo	Stars in these degrees & (Constellation)
1) AIR	01/23/2021 Aquarius 23	02/08/2021 Aquarius 21	03/06/2021 Aquarius 20	
2) AIR	05/17/2021 Gemini 19	06/10/2021 Gemini 21	07/05/2021 Gemini 23	Capella* (Auriga)
3) AIR	09/13/2021 Libra 19	10/09/2021 Libra 17 (Mars)	10/25/2021 Libra 14	Kraz (Corvus)
4) AIR	01/07/2022 Aquarius 7	01/23/2022 Aquarius 4	02/16/2022 Aquarius 3	
5) AIR	04/29/2022 Taurus 30	05/21/2022 Gemini 1	06/16/2022 Gemini 3	
6) AIR	08/27/2022 Libra 2	09/23/2022 Libra 1	10/08/2022 Virgo 28	

Dates from **01/07/2023** to **04/10/2024** belong the Mercury Elemental Years of Earth

ELEMENT ▽	Max E Eve Star Above W Horizon Separation/ Nigredo	Conjunction Below Horizon (other planets conj) Initiation/Albedo	Max E Morn Star Above E Horizon Return/Rubedo	Stars in these degrees & (Constellation)
1) EARTH	12/21/2022 Capricorn 21	01/07/2023 Capricorn 17	01/30/2023 Capricorn 16	
2) EARTH	04/11/2023 Taurus 12	05/01/2023 Taurus 12	05/29/2023 Taurus 14	Menkar (Cetus)
3) EARTH	08/09/2023 Virgo 15	09/06/2023 Virgo 14	09/22/2023 Virgo 12	Mizar/Alcor (Ursa Major)
4) EARTH	12/04/2023 Capricorn 4	12/22/2023 Capricorn 1 (Winter Solstice)	01/12/2024 Sagittarius 29	(Galactic Equator)

Dates from **04/11/2024** to **11/19/2025** belong the Mercury Elemental Years of Fire

ELEMENT △	Max E Eve Star Above W Horizon Separation/ Nigredo	Conjunction Below Horizon (other planets conj) Initiation/Albedo	Max E Morn Star Above E Horizon Return/Rubedo	Stars in these degrees & (Constellation)
1) FIRE	03/24/2024 Aries 24	04/11/2024 Aries 23 (Chiron)	05/09/2024 Aries 24	Baten Kaitos (Cetus)
2) FIRE	07/21/2024 Leo 27	08/18/2024 Leo 27	09/04/2024 Leo 26	Alphard (Hydra)
3) FIRE	11/16/2024 Sagittarius 18	12/05/2024 Sagittarius 15	12/25/2024 Sagittarius 13	Ras Algethi (Hercules)
4) FIRE	03/07/2025 Aries 7	03/24/2025 Aries 5 (Venus)	04/21/2025 Aries 5	
5) FIRE	07/04/2025 Leo 9	07/31/2025 Leo 9	08/19/2025 Leo 9	

Dates from **11/20/2025** to **02/17/2027** belong the Mercury Elemental Years of Water

ELEMENT ▽	Max E Eve Star Above W Horizon Separation/ Nigredo	Conjunction Below Horizon (other planets conj) Initiation/Albedo	Max E Morn Star Above E Horizon Return/Rubedo	Stars in these degrees & (Constellation)
1) WATER	10/29/2025 Sagittarius 1	11/20/2025 Scorpio 29 (Moon)	12/07/2025 Scorpio 26	Toliman* (Centaurus)
2) WATER	02/19/2026 Pisces 20	03/07/2026 Pisces 17	04/03/2026 Pisces 17	
3) WATER	06/15/2026 Cancer 20	07/12/2026 Cancer 21	08/02/2026 Cancer 22	Castor* (Gemini)
4) WATER	10/11/2026 Scorpio 14	11/04/2026 Scorpio 13	11/20/2026 Scorpio 10	Alphecca* (Corona Borealis)

Dates from **02/18/2027** to **06/22/2027** belong the Mercury Elemental Year of Air

ELEMENT △	Max E Eve Star Above W Horizon Separation/ Nigredo	Conjunction Below Horizon (other planets conj) Initiation/Albedo	Max E Morn Star Above E Horizon Return/Rubedo	Stars in these degrees & (Constellation)
1) AIR	02/03/2027 Pisces 3	02/18/2027 Aquarius 30	03/17/2027 Aquarius 29	Enif (Pegasus)

Dates from **06/23/2027** to **10/18/2027** belong the Mercury Elemental Year of Water

ELEMENT ▽	Max E Eve Star Above W Horizon Separation/ Nigredo	Conjunction Below Horizon (other planets conj) Initiation/Albedo	Max E Morn Star Above E Horizon Return/Rubedo	Stars in these degrees & (Constellation)
1) WATER	05/28/2027 Cancer 1	06/23/2027 Cancer 2 (Summer Solstice)	07/15/2027 Cancer 3	

Dates from **10/19/2027** to **01/15/2029** belong the Mercury Elemental Years of Air

ELEMENT △	Max E Eve Star Above W Horizon Separation/ Nigredo	Conjunction Below Horizon (other planets conj) Initiation/Albedo	Max E Morn Star Above E Horizon Return/Rubedo	Stars in these degrees & (Constellation)
1) AIR	09/24/2027 Libra 28	10/19/2027 Libra 27	11/04/2027 Libra 24	Izar (Bootes)
2) AIR	1/17/2028 Aquarius 17	02/02/2028 Aquarius 14	02/27/2028 Aquarius 12	ALNAIR (Grus)
3) AIR	05/09/2028 Gemini 11	06/01/2028 Gemini 13 (Venus & Uranus)	06/26/2028 Gemini 15	
4) AIR	09/05/2028 Libra 12	10/02/2028 Libra 10 (Jupiter)	10/17/2028 Libra 8	Gienah* (Corvus)

Dates from **01/16/2029** to **12/14/2030** belong the Mercury Elemental Years of Earth

ELEMENT ▽	Max E Eve Star Above W Horizon Separation/ Nigredo	Conjunction Below Horizon (other planets conj) Initiation/Albedo	Max E Morn Star Above E Horizon Return/Rubedo	Stars in these degrees & (Constellation)
1) EARTH	12/31/2028 Capricorn 30	01/16/2029 Capricorn 27	02/08/2029 Capricorn 25	
2) EARTH	04/21/2029 Taurus 22	05/12/2029 Taurus 23	06/08/2029 Taurus 25	
3) EARTH	08/19/1929 Virgo 25	09/15/2029 Virgo 24	10/01/2029 Virgo 22	Alkes (Crater)
4) EARTH	12/14/2029 Capricorn 14	12/31/2029 Capricorn 11	01/22/2030 Capricorn 9	Nunki (Sagittarius)
5) EARTH	04/04/2030 Taurus 4	04/23/2030 Taurus 4	05/21/2030 Taurus 5	Caph (Cassiopeia)
6) EARTH	08/01/2030 Virgo 8	08/29/2030 Virgo 7	09/15/2030 Virgo 5	Thuban (Draco)

Dates from **12/15/2030** to **03/15/2032** belong the Mercury Elemental Years of Fire

ELEMENT △	Max E Eve Star Above W Horizon Separation/ Nigredo	Conjunction Below Horizon (other planets conj) Initiation/Albedo	Max E Morn Star Above E Horizon Return/Rubedo	Stars in these degrees & (Constellation)
1) FIRE	11/26/2030 Sagittarius 27	12/15/2030 Sagittarius 24	01/04/2031 Sagittarius 22	Ras Alhague (Ophiucus)
2) FIRE	03/18/2031 Aries 16	04/04/2031 Aries 15 (Neptune)	05/02/2031 Aries 16	Alpheratz (Andromeda)
3) FIRE	07/15/2031 Leo 20	08/11/2031 Leo 20 (Venus)	08/29/2031 Leo 19	
4) FIRE	11/09/2031 Sagittarius 11	11/29/2031 Sagittarius 8	12/18/2031 Sagittarius 6	Antares* (Scorpius)

Mercury Elemental Year Ephemeris

Dates from **03/16/2032** to **07/22/2032** belong the Mercury Elemental Year of Water

ELEMENT ▽	Max E Eve Star Above W Horizon Separation/ Nigredo	Conjunction Below Horizon (other planets conj) Initiation/Albedo	Max E Morn Star Above E Horizon Return/Rubedo	Stars in these degrees & (Constellation)
1) WATER	02/29/2032 Pisces 29	03/16/2032 Pisces 28 (Vernal Point)	04/13/2032 Pisces 28	Scheat (Pegasus)

Dates from **07/23/2032** to **11/12/2032** belong the Mercury Elemental Year of Fire

ELEMENT △	Max E Eve Star Above W Horizon Separation/ Nigredo	Conjunction Below Horizon (other planets conj) Initiation/Albedo	Max E Morn Star Above E Horizon Return/Rubedo	Stars in these degrees & (Constellation)
1) FIRE	06/25/2032 Leo 1	07/23/2032 Leo 2 (Mars)	08/11/2032 Leo 2	

Dates from **11/13/2032** to **02/10/2034** belong the Mercury Elemental Years of Water

ELEMENT ▽	Max E Eve Star Above W Horizon Separation/ Nigredo	Conjunction Below Horizon (other planets conj) Initiation/Albedo	Max E Morn Star Above E Horizon Return/Rubedo	Stars in these degrees & (Constellation)
1) WATER	10/21/2032 Scorpio 24	11/13/2032 Scorpio 22	11/30/2032 Scorpio 19	
2) WATER	02/12/2033 Pisces 13	02/27/2033 Pisces 10	03/26/2033 Pisces 9	
3) WATER	06/07/2033 Cancer 12	07/04/2033 Cancer 13 (Saturn)	07/25/2033 Cancer 14	Sirius* (Canis Major)
4) WATER	10/04/2033 Scorpio 8	10/28/2033 Scorpio 6	11/13/2033 Scorpio 3	

Dates from **02/11/2034** to **01/09/2036** belong the Mercury Elemental Years of Air

ELEMENT △	Max E Eve Star Above W Horizon Separation/ Nigredo	Conjunction Below Horizon (other planets conj) Initiation/Albedo	Max E Morn Star Above E Horizon Return/Rubedo	Stars in these degrees & (Constellation)
1) AIR	01/26/2034 Aquarius 26	02/11/2034 Aquarius 23 (Pluto)	03/09/2034 Aquarius 22	Deneb Algedi (Capricornus) Sadalsuud (Aquarius)
2) AIR	05/20/2034 Gemini 22	06/14/2034 Gemini 24	07/08/2034 Gemini 26	Alnilam (Orion)
3) AIR	09/16/2034 Libra 21	10/12/2034 Libra 20 (Moon)	10/27/2034 Libra 17	
4) AIR	01/10/2035 Aquarius 10	01/26/2035 Aquarius 7	02/19/2035 Aquarius 5	
5) AIR	05/02/2035 Gemini 3	05/24/2035 Gemini 4	06/19/2035 Gemini 7	
6) AIR	08/30/2035 Libra 5	09/25/2035 Libra 3 (Autumnal Equinox)	10/11/2035 Libra 1	

Dates from **01/10/2036** to **04/13/2037** belong the Mercury Elemental Years of Earth

ELEMENT ▽	Max E Eve Star Above W Horizon Separation/ Nigredo	Conjunction Below Horizon (other planets conj) Initiation/Albedo	Max E Morn Star Above E Horizon Return/Rubedo	Stars in these degrees & (Constellation)
1) EARTH	12/24/2035 Capricorn 23	01/10/2036 Capricorn 20	02/02/36 Capricorn 19	
2) EARTH	04/13/2036 Taurus 15	05/04/2036 Taurus 15	05/31/2036 Taurus 17	Menkar (Cetus)
3) EARTH	08/11/2036 Virgo 18	09/08/2036 Virgo 17 (Mars)	09/24/2036 Virgo 15	
4) EARTH	12/06/2036 Capricorn 7	12/24/2036 Capricorn 4	01/14/2037 Capricorn 2	

Dates from **04/14/2037** to **03/09/2039** belong the Mercury Elemental Years of Fire

ELEMENT △	Max E Eve Star Above W Horizon Separation/ Nigredo	Conjunction Below Horizon (other planets conj) Initiation/Albedo	Max E Morn Star Above E Horizon Return/Rubedo	Stars in these degrees & (Constellation)
1) FIRE	03/27/2037 Aries 27	04/14/2037 Aries 26 (Neptune)	05/12/2037 Aries 27	
2) FIRE	07/25/2037 Leo 30	08/21/2037 Leo 30 (Saturn)	09/07/2037 Leo 28	Regulus* (Leo)
3) FIRE	11/19/2037 Sagittarius 20	12/08/2037 Sagittarius 18	12/28/2037 Sagittarius 15	Ras Algethi (Hercules)
4) FIRE	03/10/2038 Aries 9	03/27/2038 Aries 8	04/24/2038 Aries 8	
5) FIRE	07/07/2038 Leo 12	08/03/2038 Leo 12 (Jupiter)	08/22/2038 Leo 12	
6) FIRE	11/01/2038 Sagittarius 4	11/22/2038 Sagittarius 1	12/10/2038 Scorpio 29	Dschubba (Scorpius)

Dates from **03/10/2039** to **10/20/2040** belong the Mercury Elemental Years of Water

ELEMENT ▽	Max E Eve Star Above W Horizon Separation/ Nigredo	Conjunction Below Horizon (other planets conj) Initiation/Albedo	Max E Morn Star Above E Horizon Return/Rubedo	Stars in these degrees & (Constellation)
1) WATER	02/22/2039 Pisces 23	03/10/2039 Pisces 20	04/06/2039 Pisces 20	
2) WATER	06/18/2039 Cancer 23	07/16/2039 Cancer 24 (Uranus)	08/05/2039 Cancer 24	Pollux* (Gemini)
3) WATER	10/14/2039 Scorpio 17	11/07/2039 Scorpio 15	11/23/2039 Scorpio 12	Zuben Elgenubi* (Southern Scale)
4) WATER	02/05/2040 Pisces 6	02/21/2040 Pisces 3	03/18/2040 Pisces 2	Fomalhaut* (Pisces Austrinus)

5) WATER	05/30/2040 Cancer 4	06/25/2040 Cancer 5	07/14/2040 Cancer 6		

Dates from **10/21/2040** to **01/17/2042** belong the Mercury Elemental Years of Air

ELEMENT △	Max E Eve Star Above W Horizon Separation/ Nigredo	Conjunction Below Horizon (other planets conj) Initiation/Albedo	Max E Morn Star Above E Horizon Return/Rubedo	Stars in these degrees & (Constellation)
1) AIR	09/26/1940 Scorpio 1	10/21/2040 Libra 29	11/05/2040 Libra 26	
2) AIR	01/19/2041 Aquarius 19	02/03/2041 Aquarius 16	03/01/2041 Aquarius 15	ALNAIR (Grus)
3) AIR	05/12/2041 Gemini 14	06/05/2041 Gemini 16	06/30/2041 Gemini 18	Rigel* (Orion)
4) AIR	09/08/2041 Libra 14	10/05/2041 Libra 13	10/21/2041 Libra 10	Algorab (Corvus)

Dates from **01/18/2042** to **12/17/2043** belong the Mercury Elemental Years of Earth

ELEMENT ▽	Max E Eve Star Above W Horizon Separation/ Nigredo	Conjunction Below Horizon (other planets conj) Initiation/Albedo	Max E Morn Star Above E Horizon Return/Rubedo	Stars in these degrees & (Constellation)
1) EARTH	01/02/2042 Aquarius 3	01/18/2042 Capricorn 30 (Venus)	02/11/2042 Capricorn 28	ALTAIR* (Aquila)
2) EARTH	04/24/2042 Taurus 25	05/16/2042 Taurus 26	06/11/2042 Taurus 28	Algol* (Perseus)
3) EARTH	08/22/2042 Virgo 28	09/18/2042 Virgo 26	10/04/2042 Virgo 24	Alkaid* (Ursa Major)
4) EARTH	12/17/2042 Capricorn 16	01/03/2043 Capricorn 13	01/25/2043 Capricorn 12	Vega* (Lyra)
5) EARTH	04/06/2043 Taurus 7	04/26/2043 Taurus 7 (Neptune)	05/23/2043 Taurus 8	Hamal (Aries)

6) EARTH	08/04/2043 Virgo 11	09/01/2043 Virgo 10 (Venus)	09/17/2043 Virgo 8	Zosma (Leo)

Dates from **12/18/2043** to **03/18/2045** belong the Mercury Elemental Years of Fire

ELEMENT △	Max E Eve Star Above W Horizon Separation/ Nigredo	Conjunction Below Horizon (other planets conj) Initiation/Albedo	Max E Morn Star Above E Horizon Return/Rubedo	Stars in these degrees & (Constellation)
1) FIRE	11/29/2043 Sagittarius 30	12/18/2043 Sagittarius 27 (Winter Solstice)	01/07/2044 Sagittarius 25	(Galactic Center)
2) FIRE	03/20/2044 Aries 19	04/06/2044 Aries 18	05/04/2044 Aries 19	
3) FIRE	07/17/2044 Leo 23	08/14/2044 Leo 22 (Uranus)	08/31/2044 Leo 21	
4) FIRE	11/11/2044 Sagittarius 13	12/01/2044 Sagittarius 11 (Saturn)	12/20/2044 Sagittarius 9	Antares* (Scorpius)

Dates from **03/19/2045** to **07/25/2045** belong the Mercury Elemental Year of Water

ELEMENT ▽	Max E Eve Star Above W Horizon Separation/ Nigredo	Conjunction Below Horizon (other planets conj) Initiation/Albedo	Max E Morn Star Above E Horizon Return/Rubedo	Stars in these degrees & (Constellation)
1) WATER	03/03/2045 Aries 2	03/19/2045 Pisces 30 (Venus) (Vernal Point)	04/16/2045 Pisces 30	Scheat (Pegasus)

Dates from **07/26/2045** to **11/14/2045** belong the Mercury Elemental Year of Fire

ELEMENT △	Max E Eve Star Above W Horizon Separation/ Nigredo	Conjunction Below Horizon (other planets conj) Initiation/Albedo	Max E Morn Star Above E Horizon Return/Rubedo	Stars in these degrees & (Constellation)
1) FIRE	06/28/2045 Leo 4	07/26/2045 Leo 5	08/14/2045 Leo 4	

Dates from **11/15/2045** to **02/13/2047** belong the Mercury Elemental Years of Water

ELEMENT ▽	Max E Eve Star Above W Horizon Separation/ Nigredo	Conjunction Below Horizon (other planets conj) Initiation/Albedo	Max E Morn Star Above E Horizon Return/Rubedo	Stars in these degrees & (Constellation)
1) WATER	10/24/2045 Scorpio 27	11/15/2045 Scorpio 25	12/03/2045 Scorpio 22	
2) WATER	02/14/2046 Pisces 15	03/02/2046 Pisces 13	03/29/2046 Pisces 12	Ankaa (Phoenix)
3) WATER	06/10/2046 Cancer 15	07/07/2046 Cancer 16	07/28/2046 Cancer 17	Canopus* (Argo Navis)
4) WATER	10/07/2046 Scorpio 11	10/31/2046 Scorpio 9 (Venus)	11/16/2046 Scorpio 6	

Dates from **02/14/2047** to **01/10/2049** belong the Mercury Elemental Years of Air

ELEMENT △	Max E Eve Star Above W Horizon Separation/ Nigredo	Conjunction Below Horizon (other planets conj) Initiation/Albedo	Max E Morn Star Above E Horizon Return/Rubedo	Stars in these degrees & (Constellation)
1) AIR	01/29/2047 Aquarius 29	02/14/2047 Aquarius 26	03/12/2047 Aquarius 25	
2) AIR	05/23/2047 Gemini 25	06/17/2047 Gemini 27 (Mars)	07/11/2047 Gemini 29	Polaris (Ursa Minor) Betelgeuse* (Orion)
3) AIR	09/19/2047 Libra 24	10/15/2047 Libra 22	10/30/2047 Libra 19	Spica* (Virgo) Arcturus* (Bootes)
4) AIR	01/13/2048 Aquarius 12	01/28/2048 Aquarius 9	02/22/2048 Aquarius 8	
5) AIR	05/04/2048 Gemini 6	05/27/2048 Gemini 7 (Venus & Jupiter)	06/21/2048 Gemini 10	Aldebaran* (Taurus)
6) AIR	09/01/2048 Libra 8	09/27/2048 Libra 6	10/13/2048 Libra 3	

Dates from **01/11/2049** to **04/17/2050** belong the Mercury Elemental Years of Earth

ELEMENT	Max E Eve Star Above W Horizon Separation/ Nigredo	Conjunction Below Horizon (other planets conj) Initiation/Albedo	Max E Morn Star Above E Horizon Return/Rubedo	Stars in these degrees & (Constellation)
▽				
1) EARTH	12/26/2048 Capricorn 26	01/11/2049 Capricorn 23 (Saturn)	02/04/2049 Capricorn 21	Peacock (Pavo)
2) EARTH	04/16/2049 Taurus 17	05/07/2049 Taurus 18 (Neptune)	06/03/2049 Taurus 20	
3) EARTH	08/14/2049 Virgo 21	09/11/2049 Virgo 20 (Uranus)	09/26/2049 Virgo 17	Denebola (Leo)
4) EARTH	12/09/2049 Capricorn 9	12/27/2049 Capricorn 6 (Venus)	01/17/2050 Capricorn 5	

Dates from **04/18/2050** to **08/23/2050** belong the Mercury Elemental Years of Fire

ELEMENT	Max E Eve Star Above W Horizon Separation/ Nigredo	Conjunction Below Horizon (other planets conj) Initiation/Albedo	Max E Morn Star Above E Horizon Return/Rubedo	Stars in these degrees & (Constellation)
△				
1) FIRE	03/30/2050 Aries 30	04/18/2050 Aries 29	05/15/2050 Aries 30	Alrisha (Pisces)

Dates from **08/24/2050** to **12/10/2050** belong the Mercury Elemental Years of Earth

ELEMENT	Max E Eve Star Above W Horizon Separation/ Nigredo	Conjunction Below Horizon (other planets conj) Initiation/Albedo	Max E Morn Star Above E Horizon Return/Rubedo	Stars in these degrees & (Constellation)
▽				
1) EARTH	07/28/2050 Virgo 3	08/24/2050 Virgo 3	09/10/2050 Virgo 1	

Dates from **12/11/2050** to **03/11/2052** belong the Mercury Elemental Years of Fire

ELEMENT △	Max E Eve Star Above W Horizon Separation/ Nigredo	Conjunction Below Horizon (other planets conj) Initiation/Albedo	Max E Morn Star Above E Horizon Return/Rubedo	Stars in these degrees & (Constellation)
1) FIRE	11/22/2050 Sagittarius 23	12/11/2050 Sagittarius 20 (Chiron)	12/31/2050 Sagittarius 18	Ras Alhague (Hercules)
2) FIRE	03/13/2051 Aries 12	03/30/2051 Aries 11	04/27/2051 Aries 11	Alderamin (Cepheus)
3) FIRE	07/10/2051 Leo 15	08/07/2051 Leo 15 (Venus)	08/25/2051 Leo 14	Dubhe (Ursa Major)
4) FIRE	11/04/2051 Sagittarius 6	11/25/2051 Sagittarius 4	12/13/2051 Sagittarius 2	Dschubba (Scorpius)

Notes

1. Woody Allen, et al. *Midnight in Paris* [film] (Sony Pictures Home Entertainment, 2011).
2. Erin Sullivan, *Retrograde Planets: Traversing the Inner Landscape* (York Beach, ME: Samuel Weiser, 2000), 47.
3. Religious historian Mircea Eliade uses the term hierophany to denote the original manifestation or breakthrough of the sacred into the world. Further, Eliade suggests that as the alchemist raises her own spiritual awareness via the recapitulation of these hierophanies, she also assists in the transmutation of the cosmos. See, Mircea Eliade, *The Forge and the Crucible* (Chicago: University of Chicago Press, 1962), 165.
4. Robert G. Strom and Ann L. Sprague, *Exploring Mercury: The Iron Planet* (Chichester, UK: Praxis Publishing, 2003), 1.
5. I first became aware of this pattern through Erin Sullivan's *Retrograde Planets*, mentioned previously.
6. Hermes Trismegistus, *The Emerald Tablet*. Translation used here: Dennis Hauck, *The Emerald Tablet: Alchemy For Personal Transformation* (New York: Penguin, 1999), 45.
7. Arthur Schopenhauer, *The World as Will and Representation*, Vol. 2, trans. E. F. J. Payne, (New York: Dover, 1966), 391.
8. Two modern astrologers, Robert Blaschke and Arielle Guttman, apply this technique to the other planets. See: Robert P. Blaschke, *Astrology A Language of Life: Volume V Holographic Transits* (Port Townsend, WA: Earthwalk School of Astrology, 2006); and Arielle Guttman, *Venus Star Rising: A New Cosmology for the 21st Century* (Sante Fe, NM: Sophia Venus Productions, 2010).
9. See, Sam Wasson, *Fifth Avenue, 5AM: Audrey Hepburn, Breakfast at Tiffany's, and the Dawn of the Modern Woman* (New York: Harper, 2010).
10. The data for Leigh has only a B rating, which means it comes from a biographical source rather than an official record, thus it is only probable, rather than relatively certain, that the data is accurate.

11 Ervin Laszlo and Allan Combs, *Thomas Berry: Dreamer of The Earth: The Spiritual Ecology of the Father of Environmentalism* (Rochester, VT: Inner Traditions, 2011), 43.
12 Werner Herzog, *The Cave of Forgotten Dreams* [film] (IFC Films, 2010).
13 More than one researcher has come to this hypothesis. The paintings in Lascaux appear to contain elements of Taurus the Bull such as the Hyades and Pleiades star clusters, and possibly other constellations. See for instance the work of Dr. Michael Rappenglück, Chantal Jegues-Wolkiewiez and Jean-Michel Geneste.
14 Gebser distinguished the following structures: the *archaic* structure, the *magic* structure, the *mythical* structure, the *mental* structure and the *integral* structure. Jean Gebser, *The Ever-Present Origin*, trans. Noel Barstad with Algis Mickunas, (1985; repr. Athens: Ohio University Press, 1991).
15 Mircea Eliade, *The Sacred and The Profane: The Nature of Religion* (New York: Mariner Books, 1968).
16 Moke Kupihea, *Kahuna of Light: The World of Hawaiian Spirituality* (Rochester, VT: Inner Traditions, 2001).
17 "Sky astrologers" refers to a modern movement within the astrological community to renew the deepest, most ancient visual and cyclic traditions of astrology (which pre-date horoscopy), in order to help find and create meaning in our lives.
18 Christopher Vogler, *The Writer's Journey: Mythic Structure for Writers* (Studio City, CA: Michael Wiese Productions, 1998).
19 W. K. C. Guthrie, *The Greeks and Their Gods* (Boston: Beacon Press, 1950), 92–94.
20 My personal favorite resource for animal medicine is: Jamie Sams and David Carson, *Medicine Cards: The Discovery of Power Through the Ways of Animals* (New York: St. Martins Press, 1988).
21 You can look up your personal Mercury elemental year in the appendix.

22 In chapter six we will explore in depth all twelve of the returns of the mercury elemental year to the birth element which happen over the course of 79 years.
23 The ephemerides in the appendix clearly lay out the beginning, middle, and end dates for each of these three periods each year.
24 See Henry David Thoreau's 1854 book *Walden*.
25 See the facsimile edition of C. G. Jung's *The Red Book: Liber Novus*, ed. S. Shamdasani, trans. M. Kyburz, J. Peck, and S. Shamdasani, (New York: W. W. Norton, 2009).
26 Ken Wilber, *Up from Eden: A Trans-Personal View of Human Evolution* (1981; repr. Wheaton, IL: The Theosophical Publishing House, 1996), 32.
27 Hauck, *The Emerald Tablet*.
28 Joseph Campbell, *Hero with a Thousand Faces* (New York: MJF Books, 1949).
29 These three stages were called: pre-liminal, liminal, and post-liminal in, Arnold van Gennep, Monika B. Vizedon, and Gabrielle L. Caffee, *Rites of Passage* (Chicago: University of Chicago Press, 1961).
30 Lauren Artress, *Walking a Sacred Path: Rediscovering the Labyrinth as a Spiritual Tool* (New York: Riverhead Books, 1995).
31 Stanislav Grof, "Holotropic Research and Archetypal Astrology," *Archai: The Journal of Archetypal Cosmology* 1.1 (Summer 2009): 50–66.
32 Walter Burkert, *Homo Necans: The Anthropology of Ancient Greek Sacrificial Ritual and Myth*, trans. Peter Bing (Los Angeles: University of California Press, 1983), 165; see also, Apostolos N. Athanassakis, *The Homeric Hymns*, 2nd Ed. (Baltimore: Johns Hopkins University Press, 2004).
33 Murray Stein, *Jung's Map of the Soul* (Chicago: Open Court Publishing, 1998), esp. chapter five ("The Revealed and Concealed in Relations with Others," 103-24), which deals with the anima/shadow complexes.
34 See James Pennebaker, *Opening Up: The Healing Power of Expressing Emotions* (New York: Guilford Press, 1990).
35 See James Pennebaker, *Writing to Heal: A Guided Journal for Recovering from Trauma and Emotional Upheaval* (Oakland: New Harbinger, 2004).
36 Stein, *Jung's Map of the Soul*, especially chapter six ("The Way to the Deep Interior," 125-49), which deals with the *anima/us* complexes.

37 Gary Lachman, "Rene Schwaller de Lubicz and the Intelligence of the Heart," *Quest* 89.1 (January/February 2000): 4–11.
38 It began as the Goddess Astrology Podcast but after seven years and a complete cycle of Mercury's triple alignments through all four elements, I have recently re-branded it the Hermetic Astrology Podcast.
39 The exact degrees for 125 years of triple alignments are listed in the ephemerides in the appendix. The reader is advised to look ahead for any exact alignments to their personal birth chart and become prepared to invoke and activate their rare and special magic.
40 See for instance, John Martineau, *A Little Book of Coincidence in the Solar System* (New York: Walker Publishing, 2001).
41 Nathan Schwartz-Salant, *Jung on Alchemy* (Princeton, NJ: Princeton University Press, 1995), 107.
42 Stein, *Jung's Map Of The Soul*, particularly chapter eight ("Emergence of the Self [Individuation]," 171–97), which deals with the archetype of the Self and the processes of compensation and individuation.
43 Stein, *Jung's Map Of The Soul*, 176–77.
44 I have added the elements in parentheses here.
45 The use of the word "root" is interesting in that it seems Empedocles can be seen as a healer (using the roots of plants), and shaman or magician—as much as a philosopher. See, Peter Kingsley, *Ancient Philosophy, Mystery and Magic: Empedocles and Pythagorean Tradition* (Oxford: Oxford University Press, 1995)
46 "When thou hast made the quadrangle round, then is all the secret found." George Ripley, *The Compound of Alchymy; or, the Twelve Gates leading to the Discovery of the Philosopher's Stone (Liber Duodecim Portarum)*, 1471; as found in, Adam McLean, *The Alchemical Mandala: A Survey of the Mandala in Western Esoteric Traditions* (Grand Rapids, MI: Phanes Press, 2002), 56.
47 I first became aware of this cycle through Sullivan's *Retrograde Planets*, 67–76; later I realized the connection to the sequence in *The Emerald Tablet*.
48 Michael Maier, *Atalanta Fugiens, hoc est, Emblemata Nova de Secretis Naturae Chymica*. Published by Johann Theodor de Bry, 1617. Illustrated by Mathias Merian.

49 Jennifer M. Rampling, "Depicting the Medieval Alchemical Cosmos: George Ripley's Wheel of Inferior Astronomy," *Early Science and Medicine* 18.1/2 (2013), 45–86.
50 Ibid.
51 Gebser distinguished the following structures: the *archaic* structure, the *magic* structure, the *mythical* structure, the *mental* structure and the *integral* structure. See, Gebser, *The Ever-Present Origin*.
52 See for instance, *Diaphany: A Journal & Nocturne*. Inaugural Issue (2015).
53 Karl Kerenyi, *Hermes: Guide of Souls*, trans. Murray Stein (Putnam, CT: Spring Publications, 1976).
54 Erik Erikson, *Childhood and Society* (New York: Norton, 1950).
55 Ibid.
56 David Hajdu, "Forever Young? In Some Ways Yes," *New York Times*, May 24, 2011.
57 Peter Bogdanovich, *Runnin' Down a Dream* [film] (Warner Brothers & Penn Bright Entertainment, 2006).
58 Erikson, *Childhood and Society*.
59 Ibid.
60 James Hollis, *What Matters Most: Living a More Considered Life* (New York: Penguin, 2008).
61 Ibid.
62 See Stein, *Jung's Map of the Soul*.
63 Hollis, *What Matters Most*.
64 Murray Stein, *In MidLife* (Dallas, TX: Spring Publications, 1983).
65 Hollis, *What Matters Most*, 162.
66 John Malkin, "Your Own Damn Life: Michael Meade on the Story We're Born With," *The Sun* 431 (November 2011).
67 *The Taoist I Ching*, trans. Thomas Cleary, (Boston & London: Shambala, 1986)
68 See, for instance: Robert A. Johnson, *Inner Work: Using Dreams and Active Imagination for Personal Growth* (New York: HarperCollins, 1986); Joan Chodorow, *Jung on Active Imagination* (Princeton, NJ: Princeton University Press, 1997).
69 According to Astro-Databank she was born 25 December 1821 at 11:52 AM in Oxford MA, USA.

70 The data for Barton has only a B rating, which means it comes from a biographical source rather than an official record, thus it is only probable, rather than relatively certain, that the data is accurate.
71 Angeles Arrien, *The Second Half of Life: Opening the Eight Gates of Wisdom* (Boulder, CO: Sounds True, 2005), 93.
72 A. H. Maslow. *Religions, Values, and Peak Experiences* (New York: Penguin, 1964).
73 Malkin, "Your Own Damn Life."
74 Malkin, "Your Own Damn Life."
75 Michael Meade, *The Genius Myth* (Seattle: GreenFire Press, 2016).
76 Michael Meade, "The Myth of Genius, The Genius of Myth"—workshop at Pacifica Graduate Institute, July 8–10, 2016.
77 Michael Meade, "The Trouble With Genius," *The Huffington Post*, June 4, 2012. http://www.huffingtonpost.com/michael-meade-dhl/genius-fame_b_1563235.html [accessed 8 August 2016]
78 Justine Willis Toms, "Genius: The Divine Mission Of The Soul with Michael Meade," New Dimensions Radio, August 18, 2014; http://newdimensions.org/program-archive/genius-the-divine-mission-of-the-soul-with-michael-meade/ [accessed 8 August 2016]
79 C. G. Jung, *Memories, Dreams, Reflections*, ed. Aniela Jaffé, (New York: Random House, 1965).
80 Bill Moyers and Joseph Campbell, *The Power of Myth*, ed. Betty Sue Flowers, (New York: Doubleday, 1988).
81 Arthur Schopenhauer, *Parerga and Paralipomena: Short Philosophical Essays, Volume 1*, trans. E. F. J. Payne, (Oxford: Clarendon Press, 1974).
82 Lynn Bell, "A Conversation with Alexander Ruperti on Astrology's Place in the World," *The Mountain Astrologer* (Dec/Jan 1997/98).
83 For constellations, this resource is helpful: http://www.constellationsofwords.com/stars/Stars_in_longitude_order.htm
84 Editors' note: Caton has chosen an apt symbol to indicate the brighter stars, for root of the word asterisk comes from the Greek word *astēr*, or "star."

Bibliography

Allen, Woody, et al. *Midnight in Paris*. Film. Sony Pictures Home Entertainment, 2011.

Arrien, Angeles. *The Second Half of Life: Opening the Eight Gates of Wisdom*. Boulder, CO: Sounds True, 2005.

Artress, Lauren. *Walking a Sacred Path: Rediscovering the Labyrinth as a Spiritual Tool*. New York: Riverhead Books, 1995.

Athanassakis, Apostolos N. *The Homeric Hymns*, 2nd edition. Baltimore: Johns Hopkins University Press, 2004.

Bell, Lynn. "A Conversation with Alexander Ruperti on Astrology's Place in the World." *The Mountain Astrologer* (Dec/Jan 1997/98).

Blaschke, Robert P. *Astrology A Language of Life: Volume V Holographic Transits*. Port Townsend, WA: Earthwalk School of Astrology, 2006.

Bogdanovich, Peter. *Runnin' Down a Dream*. Film. Warner Brothers & Penn Bright Entertainment, 2006.

Burkert, Walter. *Homo Necans: The Anthropology of Ancient Greek Sacrificial Ritual and Myth*. Translated by Peter Bing. Los Angeles: University of California Press, 1983.

Campbell, Joseph. *Hero with a Thousand Faces*. New York: MJF Books, 1949.

Cheak, Aaron, Sabrina Dalla Valle, and Jennifer Zahrt, eds, *Diaphany: A Journal and Nocturne*. Volume 1. Auckland: Rubedo Press, 2015.

Chodorow, Joan. *Jung on Active Imagination*. Princeton, NJ: Princeton University Press, 1997.

Cleary, Thomas, trans. *The Taoist I Ching*. Boston & London: Shambhala, 1986.

Eliade, Mircea. *The Forge and the Crucible*. Chicago: University of Chicago Press, 1962.

———. *The Sacred and The Profane: The Nature of Religion*. New York: Mariner Books, 1968.

Erikson, Erik. *Childhood and Society*. New York: Norton, 1950.

Gebser, Jean. *The Ever-Present Origin*. Translation by Noel Barstad with Algis Mickunas. 1985. Repr. by Athens: Ohio University Press, 1991.

Grof, Stanislav. "Holotropic Research and Archetypal Astrology." *Archai: The Journal of Archetypal Cosmology* 1.1 (Summer 2009): 50–66.

Guthrie, W. K. C. *The Greeks and Their Gods*. Boston: Beacon Press, 1950.

Guttman, Arielle. *Venus Star Rising: A New Cosmology for the 21st Century*. Sante Fe, NM: Sophia Venus Productions, 2010.

Hajdu, David. "Forever Young? In Some Ways Yes." *New York Times*, May 24, 2011.

Hauck, Dennis. *The Emerald Tablet: Alchemy For Personal Transformation*. New York: Penguin, 1999.

Herzog, Werner. *The Cave of Forgotten Dreams*. Film. IFC Films, 2010.

Hollis, James. *What Matters Most: Living a More Considered Life*. New York: Penguin, 2008.

Johnson, Robert A. *Inner Work: Using Dreams and Active Imagination for Personal Growth*. New York: HarperCollins, 1986.

Jung, C. G. *Memories, Dreams, Reflections*. Edited by Aniela Jaffé. New York: Random House, 1965.

———. *The Red Book: Liber Novus*. Edited by S. Shamdasani. Translated by M. Kyburz, J. Peck, and S. Shamdasani. New York: W. W. Norton, 2009.

Kerenyi, Karl. *Hermes: Guide of Souls*. Translated by Murray Stein. Putnam, CT: Spring Publications, 1976.

Kingsley, Peter. *Ancient Philosophy, Mystery and Magic: Empedocles and Pythagorean Tradition*. Oxford: Oxford University Press, 1995.

Kupihea, Moke. *Kahuna of Light: The World of Hawaiian Spirituality*. Rochester, VT: Inner Traditions, 2001.

Lachman, Gary. "Rene Schwaller de Lubicz and the Intelligence of the Heart." *Quest* 89.1 (January/February 2000): 4–11.

Laszlo, Ervin and Allan Combs. *Thomas Berry: Dreamer of The Earth: The Spiritual Ecology of the Father of Environmentalism* (Rochester, VT: Inner Traditions, 2011), 43.

Maier, Michael. *Atalanta Fugiens, hoc est, Emblemata Nova de Secretis Naturae Chymica*. Published by Johann Theodor de Bry, 1617.

Malkin, John. "Your Own Damn Life: Michael Meade on the Story We're Born With." *The Sun* 431 (November 2011).

Martineau, John. *A Little Book of Coincidence in the Solar System*. New York: Walker Publishing, 2001.

Maslow, A. H. *Religions, Values, and Peak Experiences*. New York: Penguin, 1964.

McLean, Adam. *The Alchemical Mandala: A Survey of the Mandala in Western Esoteric Traditions*. Grand Rapids, MI: Phanes Press, 2002.

Meade, Michael. "The Trouble With Genius" *The Huffington Post*, June 4, 2012. http://www.huffingtonpost.com/michael-meade-dhl/genius-fame_b_1563235.html

———. *The Genius Myth*. Seattle: GreenFire Press, 2016.

———. "The Myth of Genius, The Genius of Myth"—workshop at Pacifica Graduate Institute, July 8–10, 2016.

Moyers, Bill, and Joseph Campbell. *The Power of Myth*. Edited by Betty Sue Flowers. New York: Doubleday, 1988.

Pennebaker, James. *Opening Up: The Healing Power of Expressing Emotions*. New York: Guilford Press, 1990.

———. *Writing to Heal: A Guided Journal for Recovering from Trauma and Emotional Upheaval*. Oakland: New Harbinger, 2004.

Rampling, Jennifer M. "Depicting the Medieval Alchemical Cosmos: George Ripley's Wheel of Inferior Astronomy." *Early Science and Medicine* 18.1/2 (2013), 45–86.

Ripley, George. *The Compound of Alchymy; or, the Twelve Gates leading to the Discovery of the Philosopher's Stone (Liber Duodecim Portarum)*. 1471.

Sams, Jamie, and David Carson. *Medicine Cards: The Discovery of Power Through the Ways of Animals*. New York: St. Martins Press, 1988.

Schopenhauer, Arthur. *The World as Will and Representation*, Vol. 2. Translated by E. F. J. Payne. New York: Dover, 1966.

———. *Parerga and Paralipomena: Short Philosophical Essays, Volume 1*. Translated by E. F. J. Payne. Oxford: Clarendon Press, 1974.

Schwartz-Salant, Nathan. *Jung on Alchemy*. Princeton, NJ: Princeton University Press, 1995.

Stein, Murray. *In MidLife*. Dallas, TX: Spring Publications, 1983.

———. *Jung's Map of the Soul*. Chicago: Open Court Publishing, 1998.

Strom, Robert G. and Ann L. Sprague. *Exploring Mercury: The Iron Planet*. Chichester, UK: Praxis Publishing, 2003.

Sullivan, Erin. *Retrograde Planets: Traversing the Inner Landscape*. York Beach, ME: Samuel Weiser, 2000.

Thoreau, Henry David. *Walden*. 1854.

Toms, Justine Willis. "Genius: The Divine Mission Of The Soul with Michael Meade." New Dimensions Radio, August 18, 2014. http://newdimensions.org/program-archive/genius-the-divine-mission-of-the-soul-with-michael-meade/

van Gennep, Arnold, Monika B. Vizedon, and Gabrielle L. Caffee. *Rites of Passage*. Chicago: University of Chicago Press, 1961.

Vogler, Christopher. *The Writer's Journey: Mythic Structure for Writers*. Studio City, CA: Michael Wiese Productions, 1998.

Wasson, Sam. *Fifth Avenue, 5 AM: Audrey Hepburn, Breakfast at Tiffany's, and the Dawn of the Modern Woman*. New York: Harper, 2010.

Wilber, Ken. *Up from Eden: A Trans-Personal View of Human Evolution*. 1981; Reprinted by Wheaton, IL: The Theosophical Publishing House, 1996.

SCRIBE SANGUINE
QUIA SANGUIS SPIRITUS

Hermetica Triptycha Volume 1 was typeset in Tiempos Text and Headline (Kris Sowersby), Miniscule (Thomas Huot-Marchand), & Solitaire MVB Pro (Mark van Bronkhorst).